Using Computational Fluid Dynamics

Using Computational Fluid Dynamics

C. T. SHAW

Department of Engineering
University of Warwick

PRENTICE HALL

New York London Toronto Sydney Tokyo Singapore

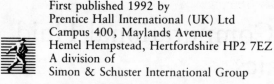

First published 1992 by
Prentice Hall International (UK) Ltd
Campus 400, Maylands Avenue
Hemel Hempstead, Hertfordshire HP2 7EZ
A division of
Simon & Schuster International Group

Typeset in $10\frac{1}{2}/12\frac{1}{2}$pt Sabon
by Mathematical Composition Setters Ltd, Salisbury, Wiltshire

Printed in Great Britain by BPCC Wheatons Ltd, Exeter

Library of Congress Cataloging-in-Publication Data

Shaw, C. T. (Chris T.)
    Using computational fluid dynamics / C.T. Shaw.
       p.   cm.
    Includes bibliographical references and index.
    ISBN 0-13-928714-0
       1. Fluid dynamics—Data processing. 2. Fluid dynamics
    —Mathematical models.   I. Title.
    QA911.S453   1992
    620.1′ 064′ 0285—dc20   91-40189
                                                            CIP

British Library Cataloguing-in-Publication Data

Shaw, C.
    Using computational fluid dynamics.
    I. Title
    620.100285

    ISBN 0-13-928714-0

1 2 3 4 5   96 95 94 93 92

To Vivienne, My Companion

The Gables

CTS 91

Contents

Contents

Contents

Figures

Figures

Tables

Preface

Computational Fluid Dynamics (CFD) can be described as the use of computers to produce information about the ways in which fluids flow in given situations. CFD embraces a variety of technologies including mathematics, computer science, engineering and physics, and these disciplines have to be brought together to provide the means of modelling fluid flows. Such modelling is used in many fields of science and engineering but, if it is to be useful, the results that it yields must be a realistic simulation of a fluid in motion. At present this depends on the problem being simulated, the software being used and the skill of the user.

Until recently the user of CFD has been a specialist, probably trained to doctoral level, working in a research and development department. Now, however, the technology is more widely available both in industry and academia and so it is being used to provide insights into many aspects of fluid motion. This increasing use has come about as there are now numerous commercial CFD software packages on the market and so it is not necessary for users to have to write their own programs in order to obtain flow results. Whilst the software is widely available, the means of learning about CFD and how to produce simulations with it tend to be restricted to post-experience courses in universities and polytechnics, where the level of assumed knowledge can be too great, or to courses run by software suppliers where users are shown how to run a particular software product. Also, there are several technical texts that describe the detailed mathematics of the modelling process, but these are often far too technical for the user of the software. Consequently, as the variety of users increases there is a need for a general text that is an introductory guide to the analysis of flow problems using CFD and describes the various stages of an analysis that must be undertaken if the user is to obtain sensible results.

Preface

This book addresses the needs of new users of CFD programs. After the introduction there is a description of some aspects of fluid flow, written specifically for the non-specialist, together with a look at some of the equations that need to be modelled. The discussion concentrates on flows which are <u>viscous and incompressible</u>, as most of the CFD packages solve this type of flow. The ways in which the governing equations are translated into a form suitable for solution by computer are then described. Having looked at this the CFD analysis process can be determined, together with some information about the software and hardware that will be required. Then each stage of the analysis process is discussed in turn, followed by a chapter where three examples of the analysis process are given. These are realistic problems which have been solved using two commercially available CFD software packages. This completes the core of the material, but as other flow types are met in practice some extensions to the basic analysis process are discussed that enable these flow types to be modelled. Then, finally, there is a review of how the necessary hardware and software can be specified. This looks at the features that might be considered, together with a discussion of how the whole process can be used to influence engineering design.

The book assumes only a minimal knowledge of fluid mechanics and mathematics, and so it is hoped that it will be a useful guide to the CFD modelling process, being read by new users of CFD software, by those interested in what CFD could do for them and even by their managers. Hopefully, the book will act both as a learning aid and as a reference. Ideally, if readers wish to perform simulations, then this book should be read in conjunction with the documentation of the appropriate software. Also, it is not intended that the book should replace the support services of the software suppliers.

For those who wish to study further, some hints on how to do this are given together with a list of key references.

C. T. Shaw

1 Introduction

1.1 Using computers to predict flows

Towards the end of 1987, two disasters occurred in Britain. In October, a severe storm swept over the South East of the country causing considerable damage and loss of life, and then, in November, there was a fire at King's Cross Underground Station in which thirty-one people died. In the investigations that followed both of these events, the use of computers to predict how fluids flow was discussed at great length.

Many people will remember that there was considerable debate as to why it was that the storm was not predicted by British weather forecasters, when forecasters in other countries did predict the storm. Forecasters use computers to predict the flow of the air in the Earth's atmosphere, finding things such as wind speed and direction, atmospheric pressure and air temperature. From this data they can predict what the weather will be several hours or days ahead. One feature of the debate was a comparison of the calculation speed and data storage capacity of the computers available to the forecasters in Britain and those more powerful machines available elsewhere. As a result of this debate, a more powerful computer has been installed in Britain for weather forecasting (Stewart, 1989). The forecasters make considerable use of the techniques known as Computational Fluid Dynamics (CFD) to produce their weather forecasts and, as we shall see later in this book, the storage capacity of the computer can affect the accuracy of the prediction, as can the speed of the machine. The results of the CFD calculations can be seen every day as part of the weather forecasts on television.

During the King's Cross fire, firemen reported that within the space of only two minutes the fire changed from being a small blaze within

1

the escalator tunnel to a serious conflagration that engulfed the booking hall at the end of the tunnel. At the inquiry that followed this disaster the results of computer predictions of the flow of air in the escalator tunnel and the booking hall were used to explain this flashover and to discount several of the theories that were put forward, such as the burning of the new paint on the ceiling of the tunnel (Highfield, 1988). These results showed a physical mechanism for the flashover, but they were so unexpected that experiments were carried out to see if such a mechanism occurred in practice. In scale models of the escalator tunnel the mechanism was found to occur, although the actual values of the flow velocity predicted by the computer were not accurate. This means that the computer predictions were correct in a qualitative sense, if not in a quantitative sense.

The timing of the two disasters and the debate that followed about the use of computers are significant. They show that from around the mid-1980s computer predictions of fluid flow have been used routinely in both science and engineering to produce useful results. The predictions have to be derived from a technology that combines advances made in several technical areas such as computer science, mathematics and engineering. These advances have contributed to the increasing use of CFD that has taken place since the above date, and it is hoped that the links between them will be seen throughout this book.

1.2 Situations where fluids flow

In many branches of engineering, there has to be an understanding of the motion of fluids. One classic example of this is in the aircraft industry, where the aerodynamics of an aircraft must be determined; i.e. the lift, drag and sideforces of a design must be estimated before a prototype flies. This ensures that the lift available will be sufficient to carry the weight of the loaded aircraft, that the required power of the engines can be determined together with the aircraft's fuel economy and that the motion of the aircraft can be predicted. To obtain this aerodynamic data many models of the design could be built and tested in a wind tunnel, with the model positioned in many orientations to the flow. Such tests might consume many hours of wind tunnel time and cost many thousands or millions of pounds.

As the equations that govern fluid motion are known, numerical approximations can be made to these equations, and, with the arrival of powerful computer hardware and software, some of the aerodynamics estimations can be made using these computer tools. This does not mean that wind tunnels are redundant. In reality, when computers and experiments are both used to produce predictions, engineers often choose to reduce the amount of wind tunnel time. Sometimes, however, the wind tunnels are used just as much as they would have been if they had been used alone. In both of these cases, wind tunnels can be used to investigate the problems that are too difficult to solve with the computational techniques, and there are many such problems. Effectively, the use of computers releases wind tunnel time and this can be used to investigate the really difficult aerodynamics problems that could not be tackled before.

Whilst this combination of experimental and computational investigations has been used to determine an aircraft's aerodynamics for some time, the use of computers for fluid flow prediction in other industrial areas is less advanced. Recently, however, other industries have been making the transition from purely experimental investigations to a mix of experimental and computational investigations. If we look at a variety of industrial sectors, such as aerospace, defence, power, process, automotive, electrical and civil engineering, there are many examples of areas where CFD is now used. For example, predictions can be made of the following:

- *Lift and drag of aircraft.* Here, as we have said, engineers need the data for performance prediction. CFD is used in conjunction with wind tunnel tests to determine the performance of various configurations.
- *Flows over missiles.* This, again, is an area where there is a need for lift, drag and sideforce data, so that simulations of performance can be made. As with aircraft, CFD and wind tunnel tests are used, but because of the wide range of flows that have to be simulated for a given configuration, use is also made of semi-empirical methods which are derived from large amounts of experimental data.
- *Jet flows inside nuclear reactor halls.* Such problems involve the simulation of fault conditions, and so engineers have great difficulty in performing actual experiments, for safety reasons. Hence, computation is the only way of trying to understand such flows.
- *Flames in burners.* There is a need to understand the complex interactions between fluid flow and chemical reaction in flames. This can assist in the production of more

3

efficient designs for burners in boilers, furnaces and other heating devices.

- *Air flow inside internal combustion engines.* When air is used to burn fuel inside an internal combustion engine, be it a gas turbine engine, a petrol engine or a diesel engine, the air must be drawn into the chamber with the minimum amount of effort, and the flow of the air once it is in the chamber must be able to promote good burning. Hence, engineers need to know the pressure drop through a system and the velocity distribution in the combustion chamber.

- *Flow of cooling air inside electrical equipment.* In this problem, electrical devices, such as integrated circuits, produce heat. This heat must be dissipated if the equipment is not to become too hot. For example, the hot devices heat the air that surrounds them and this hot air rises, creating air currents that move the heat away from the sources of heat. If insufficient heat is moved away, then it may be necessary to add fans that will force air over the hot devices.

- *Dispersion of pollutants into rivers and oceans.* Various pollutants are discharged into rivers and oceans, and computer programs can be used to predict where pollutants will travel in these naturally occurring flows and what the pollutant concentration will be at given positions in the river or sea.

From this list, it is clear that the applications can be extremely varied in nature. Despite this, the computer predictions of the different problems can be made with computer software and hardware that is not specific to a given problem. Now that these computer tools are widely available, CFD has been brought out of the research laboratory and is used by many more people. It can even be used in the engineering design process.

It is intended that this book should assist scientists and engineers in understanding how software tools can be used to predict the motion of fluids in a wide variety of situations. The emphasis is, however, on engineering examples where the speed of the flow is low and the fluid is viscous, but where the flow does not include any heat transfer. This type of flow is very common throughout industry and it can be used as the basic model upon which can be built a number of modifications that account for other types of flow. For example, the flow speed might be such that the density of the fluid will change, or heat transfer or combustion might occur.

1.3 Why read this book?

Over the last few years, many commercial CFD packages have become available. The emergence of these packages has meant that CFD is no longer practised solely in a research environment by highly trained specialists, but it is also being used in many industrial organizations as a design tool. Consequently, engineers who are not specialists in the CFD field are having to come to terms with this technology, if only in an attempt to understand what the benefits of using the technology are, and also to understand what the drawbacks are.

As a subject, CFD can appear to be far removed from the experience of those who are not specialists in the field. The situation is not helped by the numerous books on the market that address the subject of CFD, which are mainly written for the theoretical engineer or applied mathematician who is interested in the details of how the equations that govern fluid flow are solved. No general text is available for the less-specialized user of CFD techniques, or even for their managers.

There is a wide variety of people that have a need to be able to understand something about CFD techniques, be they computational analysts using CFD for the first time, design engineers interested in obtaining information about fluid motion, and even engineering managers or computer managers who provide the computational resources for CFD. Such people are invariably graduates, often with no formal background in CFD, or even in basic fluid mechanics. If these people are offered some sympathetic help and guidance, then they can understand the basics of CFD. It is the author's experience that undergraduate engineering students can successfully model fluid flow situations, if they are given appropriate background information as to what the CFD solution process is and how it is used to obtain predictions of the behaviour of fluids.

This book is an attempt to put the necessary information into a simple and concise format, so that it can be used by students or practising engineers to assist in their understanding of the technology of CFD, regardless of the particular software package they might be using. In fact, the book should act as a primer for someone about to explore the documentation of any CFD package. Once someone is familiar with the material contained here, they should be able to produce simulations of fluid flow situations using a suitable CFD package, or be able to talk confidently with those who produce such simulations.

1.4 The objectives of the study

As we have seen, CFD can be used to produce predictions for a wide variety of flows. So that the basics of the subject can be clearly understood, particularly by those outside the aircraft industry, the content of this book has in the main been restricted to the class of problems that can be described as being viscous, incompressible flows. These flows are slow speed flows where the fluid is not compressed and features such as shock waves do not occur. Many industrial flow problems are of this type, and so most of the available CFD packages can simulate these flows. There is a separate chapter that describes how to model variations from this basic type of flow.

After reading this book, it is hoped that you will be able to:

- understand something of how incompressible, viscous flows behave;
- understand the numerical techniques that are used to solve the governing equations of fluid flow;
- follow the stages undertaken during a CFD analysis;
- recognize the need for a mesh of points to be specified within the fluid volume;
- specify a flow, in terms of the relevant boundary and initial conditions;
- understand the documentation for commercial CFD software packages;
- be aware of the limitations of the CFD process.

Once the reader has this information, it should not be difficult to run some simple examples and hence gain experience in using commercial CFD packages. Having done this, the prediction of more involved fluid flow situations, where such things as heat flow, combustion and compressibility occur, should be relatively straightforward.

1.5 Using the book

The book is intended to be an introductory guide to CFD, as well as a working reference for analysts and their managers. Consequently, as readers will probably come from a variety of technical backgrounds, very little background knowledge is assumed and the book has been structured so that its chapters can be read in isolation.

Chapter 2 describes the properties of fluids that are considered important when fluids flow and describes some of the flow features that usually occur. It also provides a review of the equations that govern fluid flow and the factors that determine the flow types. This chapter is intended to be read by those with little or no formal training in fluid dynamics, and so can be skipped by other readers.

As the equations describing the flow of a fluid are partial differential equations, Chapter 3 looks at the standard ways of solving these equations using numerical approximations. Three different techniques for transforming partial differential equations into a numerical form are explained and the features ccmmon to them are emphasized. Solving the fluid flow equations leads to some special problems, regardless of the numerical technique, and so these problems and the ways of overcoming them are also explained. By using one of these techniques of approximating partial differential equations, equations can be derived which can then be programmed into a CFD software package. There is a set of operations that needs to be carried out to use such a package in a way that will produce sensible simulations of fluid flow problems, and so Chapter 4 outlines this CFD analysis process and looks at the hardware and software that are available to assist in this process.

In both Chapters 3 and 4, emphasis is given to the fact that the basic features of the software and hardware tools are common to all the packages. These chapters should be read by those who are unfamiliar with the numerical solution of partial differential equations and the software and hardware associated with such solutions.

Whilst the first four chapters cover some background material, the subsequent chapters, 5 through 9, concentrate on the CFD analysis process itself. These chapters describe in detail each of the processes that must be undertaken in order that the simulation of a fluid flow problem is successful. These processes include the formulation of the fluid flow problem, producing a flow specification that is easily translated into terms understood by the software packages, the production of a computer model, the running of the numerical solution so that reasonable results are obtained and the analysis of the results. Whilst any individual chapter forms a stand-alone module describing one particular phase in the overall process, the five chapters taken as a whole detail the analysis process from start to finish.

Having explained the analysis process in Chapters 5 through 9, Chapter 10 attempts to bring the process to life by applying the techniques described to a series of representative flow examples. It is in this chapter that we show how the techniques are actually used in practice, as the simulation process used to model these examples is described in full based on the use of commercial CFD software. From these examples the areas where CFD can be useful and the areas where it is of little use can be seen.

Finally, the last two chapters round off our study by taking a brief look, in Chapter 11, at how some of the more complex flow features such as compressibility and heat transfer are accounted for in a simulation, and then by considering, in Chapter 12, the problem of how to acquire CFD software and hardware in industry and how to implement the technology within the design process.

2 Fluids in motion

2.1 Some common flow features

When people use computers they can become so engrossed in the computational aspects of their work that everything else is excluded. For people who use CFD in an industrial environment this can be a disastrous mistake, as the computer hardware and software are merely tools to assist our understanding of the ways in which fluids flow and of the interaction between this flow and some object that is being or has been designed. Consequently, it is very important that everyone concerned with CFD has some understanding of the physical phenomena that occur when fluids flow, as it is these phenomena that CFD must analyse or predict. As this is a book that has been designed to help explain some of the mysteries of how we can predict the motion of fluids using computer-based tools, we must start by looking at the basic processes of fluid flow. These can be extremely complex and the computer simulation has to be capable of reproducing this complexity. If analysts are aware of these physical realities, they can modify their modelling technique to ensure that the best possible results for a given situation can be produced.

Whilst many engineers will have studied fluid mechanics as part of their formal education, some readers may not have made such a study, and so this chapter attempts to provide some information for those who have no formal background in the subject and for those who may wish simply to be reminded. The presentation of the material is based initially around the features that occur when fluids are flowing; that is, it considers what happens to a fluid in motion, and thereby aims to develop an intuitive feeling for the subject. Then some of the mathematical aspects of the analysis of fluids in motion are discussed. This is not intended to be a comprehensive review, but

9

it should highlight some of the more important features, giving a base for further study.

2.1.1 *Fluids all around us*

When starting to think about the way fluids flow, many people are put off by the complexity of the subject. Even the titles of the categories by which flows are classified require some knowledge of fluid flow if they are to be understood. If you look at some of the many textbooks concerned with fluid mechanics it is clear that there are many such categories and these include:

- viscous or inviscid flows;
- incompressible or compressible flows;
- flows in pipes or open channels;
- flows in pumps and turbines;
- water waves.

The relevance of some of these classifications will become clearer as we progress, but it is sufficient to note here that these do serve a useful purpose in identifying the types of flow that can be found. It is, however, just as important for someone involved with CFD to recognize the phenomena that occur for each flow type, as well as the classifications themselves.

We are going to be concerned predominantly with the use of computers to simulate flows that are found in industrial situations, outside the main stream of aeronautical applications. In many of these industrial flows the fluid moves at a low speed and so the stickiness, or viscosity, of the fluid produces forces which dominate the flow. This is especially true when the flow takes place within fixed solid boundaries. In an attempt to give a good intuitive feel for this class of flows let us consider some of the common flow features of low speed, viscous flows.

Everyone has seen many examples of the flow features that exist in industrial fluid dynamics problems. We see water coming out of a tap, litter or leaves being blown about by the wind and water flowing in rivers. By making a careful study of such things it is possible to understand a great deal about the ways in which fluids behave when they are flowing, without reading a single fluid mechanics textbook. In fact, some of the classical experiments of fluid dynamics can be re-created in the home or even experienced during a short walk.

Take, for example, the common tap by a domestic sink. Slowly turn the tap on and see that water drips out of the tap. Open the tap further to increase the flow rate until a steady column of water comes out of the tap. Notice how smooth the water column is, appearing crystal clear like glass. Increase the flow rate further and the water column surface begins to move slowly before the whole column becomes opaque. At this final stage the water flows in a direction which is generally downwards, but if we look at one point in space in the water column the fluid seems to move in a random fashion, a so-called turbulent motion, which is superimposed on the general flow. This simple experiment with the flow out of a tap demonstrates that two main types of flow can be seen with viscous fluids: first a smooth laminar flow, for example where the water moves layer over layer giving a clear column of liquid, and a randomly fluctuating turbulent flow.

A second set of flow examples can be created with a bath of water. Run several inches of water into a bath and let the natural motion of the water decay away. Then make sure that the surface of the water is illuminated, as, when the water is in motion, shadows will be cast onto the base of the bath and these will give us some clues as to the motion of the water and so help our understanding of the flow. Now, drag various objects through the water and watch what happens. For example, put a circular cylinder such as an aerosol can into the water, with its longitudinal axis in the vertical position, and then move the cylinder along. Notice that the water moves so as to flow smoothly around the front of the cylinder, but that it does not move in a similar way at the back of the cylinder. There, the water forms into tight swirls of fluid, as shown in Figure 2.1. Repeat the same experiment with a hand. First of all straighten your fingers and

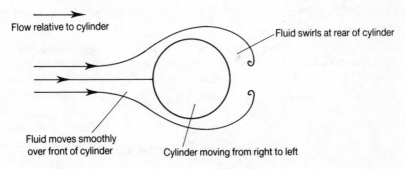

Figure 2.1. A cylinder moving in water

11

Fluids in motion

place them vertically in the water with the fingers arranged from left to right. Now move your hand to the left and see what happens. Things are much the same as for the cylinder and are shown in Figure 2.2(a). Now place your hand at a slight angle to its previous position and then move it slowly to the left. Notice the new flow

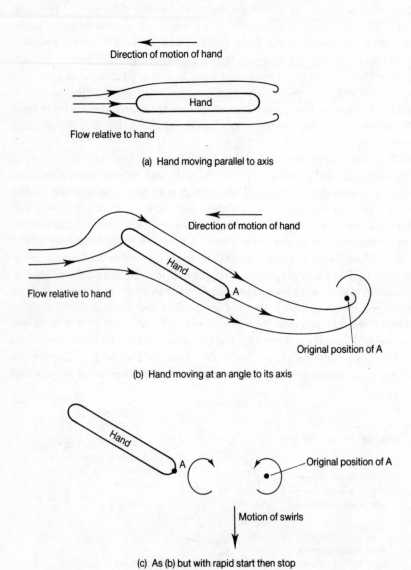

(a) Hand moving parallel to axis

(b) Hand moving at an angle to its axis

(c) As (b) but with rapid start then stop

Figure 2.2. Flow around a hand in water

12

pattern as shown in Figure 2.2(b). First, the water seems to approach your hand from below before splitting into two streams, one of which moves along the lower side of the hand and the other moves around the forward edge and then down the upper side of your hand. A swirl of fluid is left behind at the original position of the right hand end of your hand. Stop moving your hand and watch a swirl of water form which rotates in the opposite direction to the first swirl. If you perform a quick start–stop action, Figure 2.2(c), the two swirling areas of fluid move down together, as each moves under the influence of the other.

As a final experiment with the bath of water sprinkle some powder such as talc onto the water surface, and then place a sheet of card or paper in the water and drag it along so that the disturbance is a minimum. Note that the fluid nearest the card moves along with the card and appears to leave the rest of the fluid behind.

A common place where fluids flow is a river or stream, and particularly interesting effects can be seen at the point where the water flows under a bridge or around a bend. This flow will serve as our final demonstration. For example, stand on a bridge and look down into the flow. Figure 2.3 shows some of the features that can be seen. Observe that, near the bank of the river, any objects such as small insects or leaves move much more slowly in the flow than do those in the centre of the river. Looking at the diagram, near the centre of the flow an object might move from position A to position B in a given time, but near the bank an object will only move from position C to position D in the same time. Also note that near the bank,

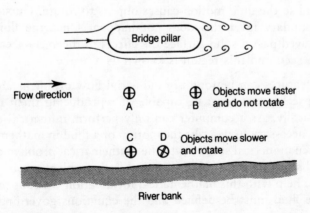

Figure 2.3. Flow in a river

objects tend to spin around, in a clockwise direction in the case shown in Figure 2.3, but that they do not spin if they are near the centre of the flow. Where there is a pillar in the water, say supporting the bridge, look at the swirling areas of fluid downstream of the pillar.

2.1.2 *The way fluids flow*

These simple demonstrations, described above, show some of the major features that are found to occur when fluids flow at slow speeds. In particular, it is important to recognize the following:

- Viscous flows can be laminar where the fluid is ordered and flows as if it was a series of sheets moving over each other.
- Viscous flows can be turbulent where the flow at one point is generally in one direction but that this mean flow has a seemingly random, fluctuating component superimposed on it.
- Normally a fluid flows cleanly around the front of an object but, around the back of an object, the direction of motion of the fluid does not stay parallel to the surface and the fluid swirls around. The fluid is said to separate from the surface and the swirls are called 'vortices'.
- When a fluid flows over a solid surface, it is slowed down by the solid surface. This is due to fluids being sticky or viscous. The area of fluid near the surface that is slowed down is called a 'boundary layer'. Inside a boundary layer the flow velocity changes with distance away from the solid surface and so the fluid motion causes objects to rotate. Outside the boundary layer this does not happen. Once the fluid has moved past a solid surface the effects of the surface can still be seen and this region is known as a 'wake'.

All these features can be found in industrial flow problems and our modelling techniques must be capable of reproducing them if they exist physically. As a computer can only perform numerical operations, it is necessary to describe the motion of a fluid in mathematical terms. Then numerical solutions to the mathematical problem can be found and so a prediction of the physical flow problem can be determined. To help with this mathematical formulation, various properties of the fluid must be defined and the equations governing their variation in both time and space developed.

14

2.1.3 *Some properties of fluids*

Fluids in motion can be described in many ways, but we need to find some way of completely describing the state of a fluid. One obvious way is to have a description of the velocity of the fluid at all points in space and time. Note that velocity is a vector quantity and so it describes both a size and a direction. One way of specifying a velocity vector is to give the components of the vector in the three Cartesian coordinate directions. This description of the velocity field does not, however, contain enough information to define the state of the fluid in full, as other properties of the fluid must be known together with the velocity. The question is now: 'Which properties do we need to describe?'

It is common knowledge that fluids can exert forces on objects. For example, in a strong wind, people and trees are blown over and slates are removed from roofs; and so the air must exert some sort of force on these objects. Forces applied by fluids are also used by a variety of means of transportation. Ships float on water as the water provides a lifting force, and aircraft fly quite successfully as the air moving over the wings also provides a lifting force. The mechanism that creates these forces is that a fluid exerts a pressure on the surface of an object, and this pressure acts in such a way that when the sum of the pressure on each small section of the surface of the object is calculated a net force exists. Pressure is the force per unit area (or stress) normal to a surface and can occur if a fluid is stationary or moving. For example, a ship floats regardless of its speed through the water, but a conventional aircraft must be moving for there to be a lifting force on its wings.

As well as this normal stress, or pressure, there is a stress derived from the action of a fluid that can act tangential to a solid surface. This stress is caused by the fact that the bulk of the fluid and the object are moving relative to each other and so the fluid is sheared. Fluids resist this shearing, such that a tangential stress acts in a direction parallel to the direction of motion of the fluid taken relative to the object. This provides a source of drag on a surface which is proportional to the viscosity, or stickiness, of the fluid. If the viscosity of the fluid is so small that it can be ignored, then the flow is said to be inviscid. This never happens in practice, but it can be a useful approximation to make when performing calculations. For the majority of the flows considered here the flow will be taken to be viscous.

15

The other major property of a fluid is its density, which is the mass of a unit volume of fluid. When we pump up a tyre the air in the tyre is compressed. This is because we force air to occupy a volume which is effectively constant and already contains some air. As there is now more mass in the same volume, the density of the air increases. For most of the situations that we will be considering we will assume that the density of the fluid does not change, which is true for low speed flows where there are no heating effects. When the density remains constant, the flow is said to be incompressible, but if the flow speed is increased to a value near that of the speed of sound in the fluid, compressibility effects become apparent. This will be dealt with in Chapter 11 as it is an additional feature that can be modelled if we make some modifications to the basic procedure that we will develop.

We have now reviewed the important properties that can be used, together with the fluid velocity, to describe the fluid flow situations that we want to model. These properties are:

- normal shear stress or pressure;
- viscosity, which enables us to find the tangential shear stress (the viscous shear stress);
- density.

If we are to calculate these properties, we must determine the mathematical relationships that govern the interaction between them. This can be done by considering some basic mechanics, as we shall now see.

2.2 Equations describing fluids in motion

Each CFD software package has to produce a prediction of the way in which a fluid will flow for a given situation. To do this the package must calculate numerical solutions to the equations that govern the flow of fluids. For the CFD analyst, therefore, it is important to have an understanding of both the basic flow features that can occur, and so must be modelled, and the equations that govern fluid flow. These equations can be found from the knowledge that the mass of fluid must be conserved, as must the momentum of the fluid. Whilst the equations will not be formally derived the underlying philosophy behind their derivation will be explained. Once these equations are known it should be a straightforward process to produce numerical predictions of all flows. This is not the case, however, as various problems arise in translating the mathematics into a numerical solution. One problem concerns the physics of the flow and how to model turbulence, as this complicates matters by having a seemingly random effect at each point in a flow. An attempt will therefore be made in this section to explain the ways in which turbulence affects a flow and how this turbulence can be modelled. Chapter 3 will look at some of the other problems concerned with the translation process.

2.2.1 *Developing the governing equations*

Whenever fluids flow the motion occurs in all three spatial dimensions, but, in an attempt to reduce the complexity of the problem, we often assume that a flow is two-dimensional. This assumption is useful as it reduces the number of variables that need to be considered, and so in this section we will consider only two-dimensional problems. Such flows contain all the features that are necessary to show the processes used to carry out the derivation of the mathematical equations, and the switch to the three-dimensional form of the equations is a straightforward extension of the processes described here.

To develop the governing equations of a flow, we consider a small part of the fluid as shown in Figure 2.4(a). Here, a rectangular, two-dimensional patch of fluid ABCD is shown, together with an assumed velocity distribution in terms of the velocity components u and v in the x- and y-directions respectively. Then, in Figure 2.4(b), we can see the forces acting in the horizontal direction on the patch of fluid caused by a normal stress σ and a shear stress τ. Note: it can be assumed that the velocity, normal stress and shear stress vary linearly across the patch of fluid, and that their values are assumed to be constant over a given edge, or face, of the patch.

First of all, for an incompressible flow, fluid cannot accumulate in the patch. This is because the fluid cannot be compressed as its density is assumed to be a constant. As a result of this incompressibility of the fluid, the total mass of fluid flowing into the patch must be zero. Across each face the mass of fluid flowing into the patch is the product of the fluid density, the area of the face and the fluid velocity normal to the face. As the density is a constant it is the same for all faces and so can be left out of the relationships for mass flow. The net mass flow is given by the sum of the masses flowing across

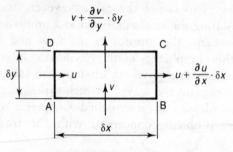

(a) Geometry and velocities

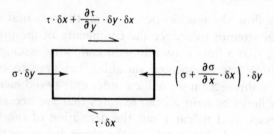

(b) Forces in x-direction

Figure 2.4. Flow into a patch of fluid

each face AB, BC, CD and DA and this is made equal to zero. Considering a positive mass flow to occur when the flow is out of the patch, this gives

$$-v\delta x+\left(u+\frac{\partial u}{\partial x}\delta x\right)\delta y+\left(v+\frac{\partial v}{\partial y}\delta y\right)\delta x-u\delta y=0 \qquad (2.1)$$

which can be rearranged to give

$$\left(\frac{\partial u}{\partial x}+\frac{\partial v}{\partial y}\right)\delta x\delta y=0$$

or just

$$\frac{\partial u}{\partial x}+\frac{\partial v}{\partial y}=0 \qquad (2.2)$$

This is known as the *continuity of mass equation*, or simply the *continuity equation*, and can be seen to be a function of the velocity components alone for an incompressible flow. If the flow is compressible the density can change and this has to be accounted for by a small modification, as we shall see in Chapter 11.

18

A second set of equations can be derived by applying Newton's Second Law of Motion to find the relationship between the forces on the patch of fluid and the acceleration of the fluid. First of all, it is necessary to determine an expression for the acceleration of the fluid that takes account of the fact that the velocity components vary in both time and space. To do this we must consider what the total change of the velocity components u or v will be due to the changes of u or v with each of the spatial directions x and y, and the time t.

Let us consider changes of the component u alone, which can be found by applying the chain rule for partial derivatives. This gives

$$\delta u = \frac{\partial u}{\partial x}\,\delta x + \frac{\partial u}{\partial y}\,\delta y + \frac{\partial u}{\partial t}\,\delta t \qquad (2.3)$$

which becomes, on dividing by δt,

$$\frac{\delta u}{\delta t} = \frac{\partial u}{\partial x}\frac{\delta x}{\delta t} + \frac{\partial u}{\partial y}\frac{\delta y}{\delta t} + \frac{\partial u}{\partial t}$$

Now, $\delta x / \delta t$ is the velocity component u itself, and similarly, $\delta y / \delta t$ is the component v, and so the relationship becomes

$$\frac{\delta u}{\delta t} = u\,\frac{\partial u}{\partial x} + v\,\frac{\partial u}{\partial y} + \frac{\partial u}{\partial t} \qquad (2.4)$$

The expression shown in equation (2.4) is the total acceleration of the fluid in the x-direction and is known as the substantive derivative of the velocity component u. It is made up of two parts: the first part consists of two terms which describe the change of the velocity component u due to the fluid being carried along, or convected, with the flow and the second part, the third term, describes the temporal change of the velocity component. When this total acceleration is multiplied by the mass of the fluid in the patch, it can be set equal to the total force in the x-direction acting on the patch of fluid. This is Newton's Second Law.

The force on the patch of fluid in the x-direction is a combination of the forces due to the normal stresses and the tangential shear stresses acting on each of the four faces of the patch. These combine to give

$$\rho\left(u\,\frac{\partial u}{\partial x} + v\,\frac{\partial u}{\partial y} + \frac{\partial u}{\partial t}\right)\delta x \delta y = \frac{\partial \sigma}{\partial x}\,\delta x \delta y + \frac{\partial \tau}{\partial y}\,\delta x \delta y \qquad (2.5)$$

where ρ is the fluid density.

Equations relating the normal stress σ to the pressure and velocity gradients and the shear stress τ to the viscosity and velocity gradients

19

Fluids in motion

can be derived (Schlichting, 1979) to give

$$\sigma = - p + 2\mu \, \frac{\partial u}{\partial x} \tag{2.6}$$

where p is the fluid pressure and μ is the viscosity, and

$$\tau = \mu \left(\frac{\partial v}{\partial x} + \frac{\partial u}{\partial y} \right) \tag{2.7}$$

When these are combined with equation (2.5) the equation that is produced in the x-direction is

$$\rho \, \frac{\partial u}{\partial t} + \rho u \, \frac{\partial u}{\partial x} + \rho v \, \frac{\partial u}{\partial y} = - \frac{\partial p}{\partial x} + \frac{\partial}{\partial x} \left(\mu \, \frac{\partial u}{\partial x} \right) + \frac{\partial}{\partial y} \left(\mu \, \frac{\partial u}{\partial y} \right) \tag{2.8}$$

and that in the y-direction is

$$\rho \, \frac{\partial v}{\partial t} + \rho u \, \frac{\partial v}{\partial x} + \rho v \, \frac{\partial v}{\partial y} = - \frac{\partial p}{\partial y} + \frac{\partial}{\partial x} \left(\mu \, \frac{\partial v}{\partial x} \right) + \frac{\partial}{\partial y} \left(\mu \, \frac{\partial v}{\partial y} \right) \tag{2.9}$$

where μ is the viscosity of the fluid and ρ is its density. Note that the effects of external forces such as gravity have been ignored here, but that they can be included as an additional force term in equations (2.5), (2.8) or (2.9) as appropriate. We will do this in Chapter 11 when we look at the effect of buoyancy on hot fluids. Also note that the viscosity μ given above is known as the dynamic viscosity, and that there is another common form of the viscosity, the kinematic viscosity ν, which is the dynamic viscosity μ divided by the density ρ.

These two equations, (2.8) and (2.9), derived from Newton's Second Law, describe the conservation of momentum in the flow and are often known as the *momentum equations* or the *Navier–Stokes* equations. They can be seen to be very similar to each other. The terms on the left hand side of each of these equations come from an acceleration term like that in equation (2.4), the second and third terms being the convection terms; whereas the right hand side terms come from the pressure gradient in the flow and the effects of viscosity.

An equation similar to the momentum equations can be derived to describe the conservation of energy within the patch, and it is this equation that is used to account for the flow of heat through a fluid, as will be described in Chapter 11.

For low speed flows without heat transfer, the equations governing the conservation of mass and momentum can be used to describe the flow exactly. That is, it should be possible to describe all incompressible flows using these equations. Turbulence, however, can make this a difficult task as, when a flow is turbulent, the velocity components

20

vary very rapidly in both space and time. Consequently, the above equations are used for laminar flows but can be used, at present, only for turbulent flows in very simple geometries such as rectangular channels. In the latter case, the amount of calculation effort required to capture both the temporal and spatial variation of the variables is extremely large, as is the amount of computer storage required to store all the necessary data for the calculation. The reasons for this will become more obvious when we look at the numerical solution of these equations in the next chapter. Most flows of interest to engineers occur in geometries which are far from simple and so, to reduce the amount of calculation effort, the turbulence has to be modelled in some simple way.

2.2.2 *Concepts of turbulence*

For those of you who carried out the experiment with the water tap that was discussed at the beginning of this chapter, we noticed that at one point in space, within the turbulent jet of water, the general fluid motion was in one direction, but that at any one point in time the flow direction was a random variation of this. Effectively, we saw a mean flow with some randomness superimposed upon it. This splitting of a flow into a mean flow and some random fluctuation gives us a guide as to how to we can model a turbulent flow. Most engineering models of turbulent flow assume that the velocity at a given point in space and a given time can be made up of the superposition of some mean velocity, which may vary slowly with time, and a random component which varies rapidly. Mathematically, the instantaneous velocity component u can be described as

$$u = \overline{U} + u' \tag{2.10}$$

where $\overline{U}$ is the mean velocity and u' is the random fluctuating component. Substituting this, and the equivalent expression for the second velocity component v, into the continuity equation (2.2), and then integrating with time gives

$$\frac{\partial \overline{U}}{\partial x} + \frac{\partial \overline{V}}{\partial y} = 0 \tag{2.11}$$

which is a time-averaged form of the continuity equation (2.2).

This simplification arises because the fluctuating components are random and so do not show any preferential direction, hence the integrals of these fluctuating components over time must be zero.

Making a similar substitution into the momentum equations (2.8)

↳Navier Stoke's equations

21

and (2.9) does not produce such a convenient result. The convection terms are non-linear terms; that is, they are the products of velocity components and the derivatives of velocity components. When we substitute expressions like the one given in equation (2.10) into the momentum equations, the convection terms generate terms for some of the products of the fluctuating components and the integral over time of these products is not zero. For example, the momentum equation in the *x*-direction, equation (2.8), becomes

$$\rho \frac{\partial \bar{U}}{\partial t} + \rho \bar{U} \frac{\partial \bar{U}}{\partial x} + \rho \bar{V} \frac{\partial \bar{U}}{\partial y}$$

$$= -\frac{\partial p}{\partial x} + \frac{\partial}{\partial x} \left(\mu \frac{\partial \bar{U}}{\partial x} \right) + \frac{\partial}{\partial y} \left(\mu \frac{\partial \bar{U}}{\partial y} \right) - \rho \frac{\overline{\partial u'^2}}{\partial x} - \rho \frac{\overline{\partial (u'v')}}{\partial y}$$

$$(2.12)$$

where the additional terms can be seen. These additional terms, which are the last two terms on the right hand side of equation (2.12) and the corresponding terms derived from substitutions into the other momentum equations, are known as *Reynolds stresses*. If we ignore these Reynolds stress terms, the time-averaged momentum equations such as equation (2.12) are the same as the original momentum equations (2.8) and (2.9), with the mean flow quantities now being substituted for the instantaneous quantities in the original equations. *It is these additional terms that are modelled to account for the effects of turbulence.*

2.2.3 *Modelling turbulence*

From our observations of turbulent flows it is clear that these flows are extremely complex. This is reflected in the increased complexity of the turbulent flow equations such as equation (2.12) where the additional terms, the Reynolds stresses, appear. When modelling these terms we try to produce simple relationships such that the final form of the equations that we solve using numerical methods is a simplification of the full equations. This means that the simplifications that are made can be so large that we reduce the accuracy of the mathematical models which provide a description of the flow. Several books describe the ways that these approximations can be made when solving engineering flow problems (Cebeci and Bradshaw, 1977; Bradshaw *et al.*, 1981), and Abbott and Basco (1989) give a comprehensive review of turbulence modelling and CFD. As a starting point these books are excellent texts.

One way of simplifying the equations is to treat the additional terms as additional viscous stresses produced by the turbulence in the flow. To do this, the Reynolds stresses are assumed to have a form similar to the viscous stresses in the momentum equations, hence the name Reynolds stress. If we consider equation (2.12), the Reynolds stress terms can be described as

$$Reynolds\ stress = \frac{\partial}{\partial x}\left(\mu_T \frac{\partial \bar{U}}{\partial x}\right) + \frac{\partial}{\partial y}\left(\mu_T \frac{\partial \bar{U}}{\partial y}\right) \tag{2.13}$$

where μ_T is an additional viscosity due to turbulence. By substituting this expression into equation (2.12) the momentum equation becomes

$$\rho \frac{\partial \bar{U}}{\partial t} + \rho \bar{U} \frac{\partial \bar{U}}{\partial x} + \rho \bar{V} \frac{\partial \bar{U}}{\partial y}$$

$$= -\frac{\partial p}{\partial x} + \frac{\partial}{\partial x}\left((\mu + \mu_T) \frac{\partial \bar{U}}{\partial x}\right) + \frac{\partial}{\partial y}\left((\mu + \mu_T) \frac{\partial \bar{U}}{\partial y}\right) \tag{2.14}$$

This equation is effectively identical to the original momentum equation (2.8), except that the mean velocity components replace the instantaneous components and the viscosity is now enhanced by an additional viscosity μ_T due to the turbulence of the flow. If this approach is followed, we can complete the modelling process if the turbulent viscosity μ_T can be found from the other flow variables. There are various ways of doing this and these include the following:

- *Mixing length arguments.* An analysis of the dimensions of the variables shows that the effective turbulent viscosity μ_T divided by the density ρ has the same dimensions as a length multiplied by a velocity. Hence momentum arguments can be used to show that μ_T is a function of the flow density, a length scale in the flow and the local mean flow velocity. Looking at equation (2.7), we see an expression for the shear stress τ which can be used to obtain the form of an expression for the turbulent viscosity. Typically, this relationship is given as

$$\mu_T = \rho c_\mu l^2 \left(\frac{\partial \bar{U}}{\partial y} + \frac{\partial \bar{V}}{\partial x}\right) \tag{2.15}$$

where c_μ is some constant that needs to be determined together with the length scale l. A numerical value for c_μ and the variation of the length scale l can be found by carrying out experiments for various simple turbulent flows such as the flow between parallel plates and the flow in pipes.

Fluids in motion

These experiments involve measuring the velocity components, pressure, laminar viscosity and density throughout the flow, and then using the momentum equations such as equation (2.14) to find the effective turbulent viscosity as a function of position. Then equation (2.15) can be used to produce values of c_μ and l by considering numerous positions in the flow.

- *Simple partial differential equation models.* Equations similar to the momentum equations can be derived that describe the distribution of the turbulent kinetic energy k which is defined for two-dimensional flows as

$$k = \tfrac{1}{2}(u'^2 + v'^2) \tag{2.16}$$

and the distribution of the dissipation rate of k, $\dot{k}$, denoted commonly by ε. As these equations describe how the variables vary throughout the field due to diffusion and convection they are known as *transport equations.* These equations are complex partial differential equations, but some of the terms in the equations are often replaced by constants which have to be found from experiments. By doing this the equations can be simplified considerably. If the turbulent kinetic energy k is found by solving the simplified transport equation, the additional turbulent viscosity can be found from (Abbott and Basco, 1989)

$$\mu_T = \rho c_\mu k^{1/2} l \tag{2.17}$$

which assumes that the mixing length l is known. The value of l might be known from experiments and, if it is known, then only the equation for k needs to be solved. This method is, therefore, known as a one-equation turbulence model. If a value for l is not known for the flow being considered, then the approximate equation for the dissipation rate ε can be solved and the additional turbulent viscosity found from (Abbott and Basco, 1989)

$$\mu_T = \rho c_\mu \frac{k^2}{\varepsilon} \tag{2.18}$$

If both partial differential equations for the turbulence parameters k and ε are solved, then we have used what is known as a two-equation turbulence model. It is the so-called k–ε model that is commonly used for most CFD calculations even

24

though it is known to be deficient for some flow types. Some five empirically derived constants are used with this model.

Another modelling approach is to try and find values for the Reynolds stresses themselves. Again, complex transport equations for these stresses have to be derived and solved. The advantage of doing this over the methods mentioned previously is that those methods give a single additional viscosity, whereas the direct modelling of the stress terms allows the effects of turbulence to vary in the three coordinate directions. It is this three-dimensional variation that is found when the stresses are measured experimentally. One- and two-equation turbulence models are said to give *isotropic turbulence*, which is turbulence which is constant in all directions, whereas in the real situation the turbulence is said to be *anisotropic*.

The two commonest ways of modelling the stresses directly are as follows:

- *Algebraic stress models*. These use a much simplified, algebraic form of the transport equations to describe the Reynolds stresses.
- *Reynolds stress models*. These use the complete form of the transport equations for the Reynolds stresses.

For the sake of completeness, we mention here the other modelling techniques that are used to model turbulent flow. These are at present only used for flows in simple geometries, and the techniques include:

- *Direct simulation*. This involves the solution of the continuity equation and the momentum equations in their simplest form, that is, equations (2.2), (2.8) and (2.9). When this is done such that the rapid variation in the variables can be determined, then there is no need for a turbulence model.
- *Large eddy simulation*. This is very similar to direct simulation, but a simple turbulence model is used to account for the very small vortices and eddies that cannot be modelled due to a lack of spatial resolution in the numerical model.

2.3 Obtaining greater understanding of fluid flow

This chapter has provided some background to the motion of fluids and the ways in which the motion can be described mathematically. For some readers the description here will be sufficient, but others will, hopefully, want to continue their study.

One of the best ways of increasing your insight into the motion of fluids is to watch fluids in motion and to observe what actually happens when fluids flow. We have seen some examples of this already and there are many more examples easily to hand. A large collection of photographs of fluids in motion has been collected and produced in one volume (see van Dyke, 1982). This is an excellent source of information as many flow features can be seen clearly. After reading this chapter, browsing through the photographs in the album should reinforce the discussion of flow phenomena that we have already made. The photographs are also very enlightening and aesthetically pleasing in their own right.

Another way of gathering information is to explore some of the many textbooks that cover the subject area of fluid mechanics. These tend to be academic texts and they lead the reader through the mathematics that describe the flow of fluids by splitting the subject into application areas. When reading the simpler material, the concepts behind fluid motion and the phenomena that occur should, by now, be more digestible. Amongst the more readable texts are those by Duncan *et al.* (1970), Goldstein (1965) and Douglas *et al.* (1985), but excellent texts of a more detailed nature are those by:

- Schlichting (1979), which deals with boundary layers and viscous flows in general;
- Bradshaw (1971), which gives a good introduction to the physics of turbulence;
- Abbott and Basco (1989), which gives a good survey of turbulence modelling;
- Hinze (1975), which gives a detailed account of the mathematics of turbulence.

For those who prefer to participate whilst learning, many short courses of instruction in fluid dynamics, aerodynamics and even computational fluid dynamics are given by higher education establishments. Many of these courses are designed specifically for people in industry and should include not only lectures but also practical sessions, where the motion of fluids can be investigated, either computationally or experimentally. Your local university or polytechnic should know the location of the centres of expertise that are close to you. Whatever you decide to do, keep your eyes and minds open, as you never know what there is of interest just around the corner.

3 Numerical solutions to partial differential equations

We have seen in Chapter 2 that the equations governing the motion of fluids are partial differential equations. These equations are made up of combinations of the flow variables, such as the velocity components and the fluid pressure, and the derivatives of these variables. Digital computers cannot be used directly to produce a solution to these partial differential equations. This is due to the fact that computers can only recognize and manipulate data in the form of zeros and ones, i.e. binary data. They can, however, be programmed to store numbers, to perform simple arithmetical operations, such as adding, subtracting, dividing and multiplying, and to repeat whole sequences of these operations on the stored numbers. Consequently, the partial differential equations have to be transformed into equations that contain only numbers, the combination of these numbers being described by the simple operations.

Producing the transformation of a partial differential equation to what is known as a numerical analogue of the equation is called *numerical discretization*. In this discretization process each term within a partial differential equation must be translated into a numerical analogue that the computer can be programmed to calculate. A variety of techniques can be used to perform this numerical discretization and, whilst each technique is based on a different set of principles, there are many common features in the methods that are used.

In this chapter we will discuss the background to three of the major numerical discretization techniques: the finite difference method, the finite element method and the finite volume method. Each of these methods will then be used to transform a simple partial differential equation into its numerical analogue. From this simple example some of the common features of the three methods and the differences between the methods can be illustrated.

Having produced a numerical analogue of a partial differential equation, the numerical equations must be processed by the computer to give a solution. This solution is a description of the magnitude of the flow variables throughout the flow field. The means of obtaining a solution to a general numerical analogue will therefore be discussed, followed by a look at the special problems that occur when we solve the numerical equations derived from the partial differential equations that govern fluid flow. It is these problems that have prevented CFD techniques from being adopted as widely as the computational techniques used to calculate the stresses and strains within structures.

As complete textbooks have been written about numerical discretization techniques and the solution of the numerical equations, it is impossible to cover all the subtle points in one chapter. This chapter should, therefore, be used as a summary of the main ideas that are used in numerical discretization, bearing in mind that the aim of this chapter is to impart some understanding of the techniques that are used to enable a computer to produce a prediction of the behaviour of a fluid. There are many sources that can be consulted if you want to study any particular aspect of this subject in more depth, and several of these are cited in the text.

3.1 Techniques of numerical discretization

3.1.1 *The finite difference method*

The first technique that we will study is known as the finite difference method. This method is based upon the use of so-called Taylor series to build a library or toolkit of equations that describe the derivatives of a variable as the differences between values of the variable at various points in space or time. A comprehensive reference to the finite difference method is Smith (1985).

When dealing with flow problems the partial differential equations discussed in Chapter 2 show us that the dependent variables are variables such as the velocity components or the fluid pressure, and that the independent variables are the spatial coordinates and time. Imagine that we know the value of some dependent variable, and all of its derivatives with respect to one independent variable, at some given value of this independent variable, a reference value. Taylor series expansions can then be used to determine the value of the dependent variable at a value of the independent variable a small distance from the reference value. For example, looking at Figure 3.1, the dependent variable U varies with the independent variable, the distance x. We can now consider the two points a small distance h away from the central point. These points are situated at $(x+h)$ and $(x-h)$ along the x-axis, and the Taylor series expansions for the variable U at the two points are

$$U(x+h) = U(x) + h\frac{dU}{dx} + \tfrac{1}{2}h^2\frac{d^2U}{dx^2} + \tfrac{1}{6}h^3\frac{d^3U}{dx^3} + \cdots \qquad (3.1)$$

and

$$U(x-h) = U(x) - h\frac{dU}{dx} + \tfrac{1}{2}h^2\frac{d^2U}{dx^2} - \tfrac{1}{6}h^3\frac{d^3U}{dx^3} + \cdots \qquad (3.2)$$

where h is the small displacement in the x-direction, and the derivatives of U are taken at the point x.

By adding or subtracting these two equations, new equations can be found for the first and second derivatives respectively at the central position x. These derivatives are

$$\frac{d^2U}{dx^2} = \frac{1}{h^2}\left(U(x+h) - 2U(x) + U(x-h)\right) + O(h^2) \qquad (3.3)$$

and

$$\frac{dU}{dx} = \frac{1}{2h}\left(U(x+h) - U(x-h)\right) + O(h^2) \qquad (3.4)$$

where $O(h^n)$ denotes that terms of order n or higher-order terms

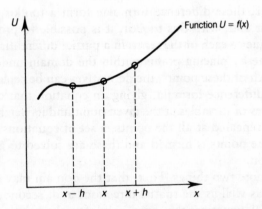

Figure 3.1. Location of points for Taylor series

29

exist. In practice, as the distance h should be small, these terms should be very small and so they will be ignored. Note that ignoring these terms leads to a source of error in the numerical calculations as the equation for the derivatives is truncated.

Further derivatives can also be formed by considering equations (3.1) and (3.2) in isolation. Looking at equation (3.1), the first-order derivative can be formed as

$$\frac{dU}{dx} = \frac{1}{h}\left(U(x+h) - U(x)\right) + O(h) \tag{3.5}$$

and similarly, from equation (3.2) another first-order derivative can be formed, i.e.

$$\frac{dU}{dx} = \frac{1}{h}\left(U(x) - U(x-h)\right) + O(h) \tag{3.6}$$

These four expressions describe some of the derivatives of the variable U at some point x by the values of the variable at the point itself, a point just behind it and a point just ahead of it, as shown in Figure 3.1. These expressions are known as difference formulae, as they involve calculating derivatives using the simple differences between the values of the variable taken at various points. Difference formulae are classified in two ways. First, by the geometrical relationship of the points and, second, by the accuracy of the expressions. Using these classifications equations (3.3) and (3.4) are central difference formulae and are second-order accurate (i.e. the neglected terms are of order h^2 or higher). Equally, equation (3.5) is a forward difference formula and equation (3.6) is a backward difference formula. Both of these two equations are first-order accurate as the neglected terms are of order h or higher.

Taken together, these difference formulae form a toolkit for the numerical analyst and, with this toolkit, it is possible to produce a numerical analogue of each of the terms in a partial differential equation. This is done by placing points within the domain under consideration. At each of these points, the derivatives can be replaced by the appropriate difference formula, giving an equation that consists solely of the values of variables at the given point and its neighbours. If this process is repeated at all the points, a set of equations for the variables at all the points is formed and these are solved to give the numerical solution.

It is useful to note two things. First, that the domain may include a time direction as well as the spatial directions and, second, that a partial differential equation that was valid for the whole of the domain, i.e. at an infinite number of points, can be translated into a finite number of equations that give the relationships between the variables at a finite set of points in the domain.

3.1.2 *The finite element method*

The second technique to be discussed is the finite element method. In this method the domain over which the partial differential equation applies is split into a finite number of sub-domains known as elements. Over each element a simple variation of the dependent variables is assumed and this piecewise description is used to build up a picture of how the variables vary over the whole domain. Intuitively, the discretization process is more complicated than that of the finite difference method, but simple examples can be used to point out the main features of the process. A good introductory text to the finite element method is Reddy (1984), but the standard reference used by finite element practitioners is Zienkiewicz and Taylor (1989).

As a historical note, the reader should be aware that the general finite element method that we will discuss emerged from computational techniques used to predict the stress and strain in solid structures. In this area of structural engineering the finite element method is now the standard computational technique used by nearly all the commercial software packages, and so many people assume that the method is only used for the solution of such problems. Now that the method has been developed into a more general computational technique, it can be used to solve a wide variety of partial differential equations and thus is suitable for the solution of many other physical problems. This confusion has led to many books being written which have the words *Finite Element Method* in their title but which deal solely with structural problems, and so these books may have little relevance to the solution of more general problems such as those derived from the equations governing fluid flow.

Let us now consider how the finite element method is used to transform a partial differential equation into its numerical analogue. First of all let us consider the element shown in Figure 3.2. On this element the variable U is assumed to vary in a simple fashion over the length of the element. In the diagram the variation is linear, but it could equally be a quadratic or cubic variation or a variation of even higher order. If the variation is linear we can describe the value of U at any point along the element as a function of the length along

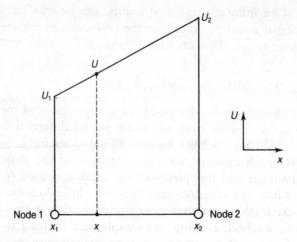

Figure 3.2. A two-noded linear element

the element x and the values of U that are known at the end-points of the element. These positions, which are used as reference positions on the element, are known as the *nodes* of the element. If the variation of the variable was assumed to be quadratic, then we would need to know the value of U at three nodes placed at, for example, the end-points of the element and the middle of the element.

With the linear variation shown, the first derivative of U with respect to x is simply a constant, and the second derivative cannot be defined. This can be a problem as many partial differential equations have terms which include second derivatives. To overcome such problems high-order derivatives can be transformed into lower-order derivatives using the following technique. First, the partial differential equation is multiplied by an unknown function, then the whole equation can be integrated over the domain in which it applies. Finally, the terms that need to have the order of their derivatives reduced are integrated by parts. This is known as producing a variational formulation.

As an example, let us consider Laplace's equation in two dimensions, where some variable Φ is described as a function of the spatial coordinates x and y. This equation is written as

$$\frac{\partial^2 \Phi}{\partial x^2} + \frac{\partial^2 \Phi}{\partial y^2} = 0 \tag{3.7}$$

To start the production of a variational formulation we multiply this by some function v and integrate it over the domain of interest

32

denoted by Ω to give

$$\int v\left(\frac{\partial^2 \Phi}{\partial x^2}+\frac{\partial^2 \Phi}{\partial y^2}\right)\, d\Omega = 0 \tag{3.8}$$

Looking at equation (3.8), each term can be seen to include second derivatives of the variable Φ and so both terms must be integrated by parts to give

$$\int \left(-\frac{\partial v}{\partial x}\frac{\partial \Phi}{\partial x}-\frac{\partial v}{\partial y}\frac{\partial \Phi}{\partial y}\right)\, d\Omega + \int \left(v\,\frac{\partial \Phi}{\partial x}\,n_x+\frac{\partial \Phi}{\partial y}\,n_y\right)\, d\Gamma = 0 \tag{3.9}$$

where Γ denotes the boundary of the domain Ω and n_x and n_y are the components of the unit outward normal vector to the boundary Γ. Note that the terms which contain the second-order derivatives in Φ have now been transformed into terms which are the products of first-order derivatives in both Φ and v. This reduction in the order of the derivatives is what we want to achieve so that a lower-order variation of the variables can be used on an element, but we can see that there is a penalty in doing this as terms on the boundary of the domain have appeared in equation (3.9), and so these must also be accounted for.

Equation (3.9) is known as the variational form of the partial differential equation (3.7) and it is this that is used to produce a discrete form of the partial differential equation for each element of the domain. The discrete form is produced by considering the variation of the variable over the element which, as we have seen, is a function of position within the element and the nodal values. We assume that the variation can be written as

$$\Phi = \sum_{i=1}^{nn} N_i \phi_i \tag{3.10}$$

where nn is the number of nodes on the element. The N_i terms are known as the shape functions and are a function of the position within the element, and the ϕ_i terms are the nodal values of Φ. For example, for the two-noded linear element shown in Figure 3.2, the shape functions can be found from the form of U, which is

$$U(x) = u_1 + \frac{(x-x_1)}{(x_2-x_1)}(u_2-u_1) \tag{3.11}$$

This can be rewritten in the form of equation (3.10) to give

$$U(x) = u_1\left(\frac{x_2-x}{x_2-x_1}\right) + u_2\left(\frac{x-x_1}{x_2-x_1}\right) \tag{3.12}$$

33

Hence, by comparing equation (3.12) to equation (3.10), the shape functions can be seen to be

$$N_1 = \frac{x_2 - x}{x_2 - x_1} \qquad (3.13)$$

and

$$N_2 = \frac{x - x_1}{x_2 - x_1} \qquad (3.14)$$

Looking at these two expressions we can see that if the value of x is set to be x_1 then N_1 is unity and N_2 is zero. Similarly, if the value of x is set to x_2 then N_1 is zero and N_2 is unity. This property is an obvious consequence of the form of equation (3.10) and can be used as a check on the algebraic expressions for a shape function regardless of whether the element is in one, two or three dimensions.

Now that we know the variation of a variable over an element, the derivatives of the variable at a point can be found. For example, to approximate the first derivatives of the variable Φ, equation (3.10) can be differentiated to give

$$\frac{d\Phi}{dx} = \sum_{i=1}^{nn} \frac{dN_i}{dx} \phi_i \qquad (3.15)$$

It should be noted here that the ϕ_i terms are not differentiated as they are constants, being the values of Φ at the nodes.

At this stage we need to know how to describe the function v. If there are two nodes on an element we need to know two functions for v. This allows us to generate the same number of equations as there are unknown values on the element. In practice there are many suitable forms for v and the standard way of specifying v is to let it be the same function as the shape function for each node. If this definition of v is used the method is known as a Galerkin method, but other methods of specification for v can also be used.

Finally, the discretization is completed by substituting equation (3.10) for the variables, equations similar to equation (3.15) for the derivatives and equations similar to equations (3.13) and (3.14) for v into the variational form and then integrating to give a series of equations for the values of the variables at the nodes of the element. For every sub-domain or element in the problem, several equations will be generated, and these equations can be collected together and then solved to find a solution.

34

3.1.3 *The finite volume method*

The third, and probably the most popular, numerical discretization method used in CFD is the finite volume method. This method is similar in some ways to the finite difference method, but some implementations of it also draw on features taken from the finite element method. The finite volume method was developed specifically to solve the equations of heat transfer and fluid flow and is described in detail by Patankar (1980).

Essentially, the governing partial differential equations are converted into numerical form by a physically based transformation of the equations. For example, the momentum equations (2.8) and (2.9) can be considered as a series of fluxes into a volume of fluid, together with a source term which is the pressure gradient. The most informative way of seeing how the process works is to consider the transformation of a typical equation and we will do this in the next section.

3.2 Numerical discretization of a simple equation

To see how these three discretization techniques are used, we will consider the discretization of the time-dependent diffusion equation:

$$\frac{\partial U}{\partial t} = \frac{\partial^2 U}{\partial x^2} \tag{3.16}$$

which consists of a first derivative in the time direction t and a second derivative in the space direction x. This is a parabolic partial differential equation that can be used to model the temporal changes in the diffusion of some quantity through a medium. As an aside, there are three classifications of partial differential equations (Smith, 1985): elliptic, parabolic and hyperbolic. Equations belonging to each of these classifications behave in different ways both physically and numerically. In particular, the direction along which any changes are transmitted is different for the three types. Depending on the flow, the governing equations of fluid motion can exhibit all three classifications. For example, the incompressible Navier–Stokes equations, equations (2.8) and (2.9), are parabolic when time-dependent, as information on changes to the flow is signalled everywhere in space but only forward in time; they are elliptic when the flow speed is low and steady as the changes are signalled everywhere; but the equations become hyperbolic if the flow speed is above the speed

35

of sound in the fluid and the changes are signalled along specific directions in space.

Having said this we can see that equation (3.16) could be regarded as a model of the momentum equations that govern an incompressible, viscous flow.

3.2.1 *Using finite differences*

To solve the above equation using finite differences we must first of all decide what the domain of the problem is. For example, equation (3.16) could be a description of the diffusion of a gas into a semiconductor of a given length and this length would then be the extent of the domain in the *x*-direction. In the time direction, however, it is usual to have positive time, that is we start the time at $t = 0$, but the extent of the domain in the positive time direction is not known as the calculation could proceed for an infinite period of time. Such a domain is said to be semi-infinite. Once we know the domain we can place points within it, and it is at these points that we perform the discretization of equation (3.16). The simplest way of placing the points within the domain is shown in Figure 3.3, where we can see part of the grid of points in the *x–t* plane. Note that there is a constant spacing δ*x* or δ*t* between each of the points in both the *x*-direction and the *t*-direction. Each of the points is labelled using an *i, j* indexing system and this denotes the position of the points in the *x*- and *t*-directions.

Having produced the grid we can now choose the difference

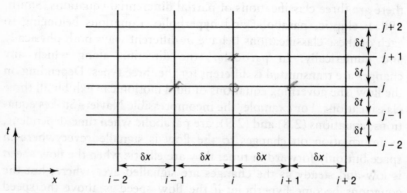

Figure 3.3. An *x–t* grid

formulae that we wish to use to produce the discrete form of equation (3.16). There are various combinations of formulae that can be used for this equation, but the simplest form of the numerical analogue is generated if we use the forward difference formula (equation (3.5)) for the time derivative that appears on the left hand side, and the central difference formula (equation (3.3)) for the spatial derivative on the right hand side. Taking the spatial derivative to be formed at the *j*th time level and to be centered on the *i*th point in *x*, and taking the time derivative to be at the *i*th *x*-position and the *j*th time level looking forward to the *j* + 1th time level, the discrete equation can be written

$$\frac{u_{i,j+1} - u_{i,j}}{\delta t} = \frac{u_{i-1,j} - 2u_{i,j} + u_{i+1,j}}{\delta x^2} \tag{3.17}$$

which can be rearranged to give

$$u_{i,j+1} = \frac{\delta t}{\delta x^2} u_{i-1,j} + \left(1 - 2\frac{\delta t}{\delta x^2}\right) u_{i,j} + \frac{\delta t}{\delta x^2} u_{i+1,j} \tag{3.18}$$

This equation may be considered to be a molecule, similar to those found in chemistry, where the four points are like atoms and are linked as shown in Figure 3.4(a). It can be clearly seen from this that the value at position $i, j+1$ depends only on the three values at the time level *j*. Consequently, if we know the values of *U* at time level *j*, the values of *U* at time level $j + 1$ are easy to calculate. To start the calculation we must, therefore, know the values of *U* at all the positions in *x* at time $t = 0$. These are known as the *initial conditions*.

Another formulation for equation (3.16) can be obtained by taking the same expression for the time derivative together with a weighted average of the spatial derivatives at the two time levels *j* and $j + 1$. This gives

$$\frac{u_{i,j+1} - u_{i,j}}{\delta t} = \theta\left(\frac{u_{i-1,j+1} - 2u_{i,j+1} + u_{i+1,j+1}}{\delta x^2}\right)$$
$$+ (1-\theta)\left(\frac{u_{i-1,j} - 2u_{i,j} + u_{i+1,j}}{\delta x^2}\right) \tag{3.19}$$

where θ and $(1 - \theta)$ are used to weight the derivatives and θ must be in the range $0 \leqslant \theta \leqslant 1.0$.

Equation (3.19) shows that there is a relationship between the three values of *U* at time level $j + 1$ and the three values of *U* at time level *j*, and so the computational molecule has changed for this case to that shown in Figure 3.4(b). Note that when θ is zero equation (3.19) is reduced to equation (3.17). When one unknown value of a

37

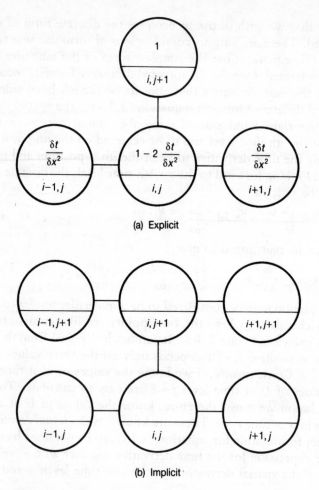

(a) Explicit

(b) Implicit

Figure 3.4. Finite difference molecules

variable can be found directly from known values of the variable, the computation is known as an *explicit* scheme (for example, equation (3.18) and Figure 3.4(a)). However, if the discretization produces an equation where several unknown values are related to several known values, for example in Figure 3.4(b) and equation (3.19) with θ not zero, then the computation is known as an *implicit* scheme. To produce a solution with an explicit scheme each unknown value of U can be easily calculated, but to produce a solution with an implicit scheme a set of simultaneous equations must be solved to find the unknown values of U.

At first sight it appears that the implicit schemes require more computational effort to produce a solution, and so we might ask ourselves the question 'Why use an implicit scheme when it involves more computational effort than an explicit scheme?' The answer to this lies in the difference in the stability of the two schemes. A stable solution is taken to be, in this case, one which progresses from time level to time level in a realistic way. An analysis of the stability (Smith, 1985) shows that for this problem, if equation (3.18) is used as the numerical analogue of the partial differential equation (3.16), then the value of the parameter $\delta t/\delta x^2$ must be less than or equal to one-half for the computational scheme to be stable. This means that where the values of δx are small the time step δt must be considerably smaller, and so with an explicit scheme there is a restriction on the size of the time step. This can mean that the time step must be very small even if the changes in the variables from one time level to the next are not very large. Implicit schemes overcome this restriction for some values of θ, and a commonly used implicit scheme takes a value of θ equal to one-half. This is known as the Crank–Nicholson scheme and is stable for all sizes of time step. Using such an implicit scheme allows a larger time step to be used than could be used with an explicit scheme, and so the computational effort for an implicit scheme can be less than that for an explicit scheme.

If we now consider the computational molecules and the grid together, it is possible to see that we still cannot solve the whole problem as we do not, as yet, have enough information. Looking at Figure 3.5 we can see an x–t grid of a domain. There are six points in the x-direction and two time levels are shown. Now let us assume that we shall use an explicit formulation, and so from the known initial conditions we can use our computational molecule to calculate the values of the variable at some points at the next time level. Given the way in which information flows from level to level, the values at the points (2,2) to (5,2) inclusive can be calculated, but we cannot use the computational molecule to find the values of the variable at the boundary points (1,2) and (6,2). To find these values we must have a knowledge of the *boundary conditions* of the problem.

For many physical problems, boundary conditions are usually given in one of the following two forms:

- *Dirichlet boundary conditions.* Here the values of the variable on the boundary are known constants. This allows a simple substitution to be made to fix the boundary value. For

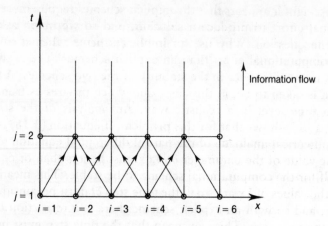

Figure 3.5. Information flow for explicit scheme

example, if U is a measure of gas concentration, we might want to assume that it is fixed at the left hand end of the domain shown in Figure 3.5, and will have a value of 10.0, say. It is easy to apply this boundary condition as we just set the value of U at the point (1,2) to 10.0.

- *Neumann boundary conditions.* Here the derivatives of the variable on the boundary are known, and this gives an extra equation which can be used to find the value at the boundary. For example, we might assume that the derivative of U is zero at the right hand end. Then if we use a first-order difference for the derivative the value of U at point (6,2) will equal the value of U at point (5,2) to satisfy this boundary condition.

Once we know both the initial conditions and the boundary conditions, we can proceed with the calculation. Using the known values at the first row of points the values of the variables at the internal points at the next row are found using an explicit scheme. Then the boundary conditions are applied to get the values at the boundary points. This gives us a second complete row of points where we know all the values of the physical variable. These can be used as a new set of initial conditions and so the process can be repeated to give the next row and so on.

With implicit schemes the handling of both fixed-value boundary conditions and derivative boundary conditions involves adding the

40

extra equations to those already generated from the partial differential equation. With these extra equations the number of equations should match the number of unknowns and so the full set of simultaneous equations can be solved.

3.2.2 *Using finite elements*

Finite element methods were originally developed to deal with steady state problems, but they can also be used to deal with time-dependent problems. We need to do this for the problem under consideration as equation (3.16) has a term which is a function of time on the left hand side. This term is dealt with first by using the forward difference formula, equation (3.5), to produce the following equation

$$\frac{U^{n+1} - U^n}{\delta t} = \frac{\partial^2 U}{\partial x^2} \tag{3.20}$$

Here, the superscripts n and $n+1$ refer to the values of U at the nth and $n+1$th time level respectively.

Now, a variational form of equation (3.20) can be produced. This is done as was shown in Section 3.1.2, by multiplying by a function v, integrating over the domain and then integrating some terms by parts, where necessary, to remove any second derivatives. This procedure gives: — *producing variational formulation*

$$\int v\left(\frac{U^{n+1} - U^n}{\delta t}\right) d\Omega = \int v\left(\frac{\partial^2 U}{\partial x^2}\right) d\Omega \tag{3.21}$$

which becomes, on integrating the second derivative on the right hand side by parts,

$$\int v\left(\frac{U^{n+1} - U^n}{\delta t}\right) d\Omega = \int \left(-\frac{\partial v}{\partial x}\frac{\partial U}{\partial x}\right) d\Omega + \int \left(v\,\frac{\partial U}{\partial x}\,n_x\right) d\Gamma$$

$$\tag{3.22}$$

This is the variational form of the equation and is also known as the *weak* form of the equation. In the original equation (3.20) the variable U had to be capable of being differentiated twice as there is a second derivative of U in the equation. Now only first derivatives of U are required, and so we say that the continuity requirement for U has been reduced from second- to first-order and is therefore weakened. This variational form must now be transformed into a numerical analogue, and this is done for a typical element of the domain. In this case the domain can be taken to be a series of lines from $x = 0$ to $x = L$ at various time levels. Hence each element is

41

effectively a one-dimensional line element similar to the one we looked at in Section 3.1.2.

Now equation (3.22) can be transformed into the numerical form using the Galerkin approach, where the multiplier v is set to be the same as the shape functions of an element. On each element the variation of U is described by:

$$U = \sum_{i=1}^{nn} N_i u_i \qquad (3.23)$$

where nn is the number of nodes on the element and the N_i terms are the shape functions, and so we can substitute for the multiplier v, for the values of U at the two time levels and for the spatial derivatives of U at the nth time level to produce an explicit form of equation (3.22). This is

$$\int N_i \left(\frac{N_j u_j^{n+1} - N_j u_j^n}{\delta t} \right) d\Omega = \int \left(-\frac{\partial N_i}{\partial x} \frac{\partial N_j u_j^n}{\partial x} \right) d\Omega$$
$$+ \int \left(v \frac{\partial U}{\partial x n_x} \right) d\Gamma \qquad (3.24)$$

Here the i, j subscripts refer to the summation in equation (3.23), and not to some position within a mesh of points, as was the case with the finite difference method example. Note that the boundary term has not been discretized, as this so-called flux can be taken to be a known value that needs to be added later. On the faces of most elements the flux term is ignored, as we assume that the fluxes cancel out across those faces that are internal to the domain. This is an equilibrium condition. It is only on the boundaries of the domain that the flux terms need to be added. If the fluxes are not added, they will be calculated by the method as being zero, and because of this they are known as *natural* boundary conditions. If we specify the value of U at a boundary then the flux term is not required, just as with the finite difference method, and this is known as an *essential* boundary condition.

For simple elements the shape functions N_i are simple functions of the coordinates, say x, and so equation (3.24) can be integrated exactly over each element, but for more complex elements this integration has to be performed numerically. If we use simple one-dimensional elements that have two nodes, as we did in Section 3.1.2, then the above equation can be integrated to yield two separate equations for each element in terms of the nodal values of U at the $n+1$th time level, if the values at time level n are known.

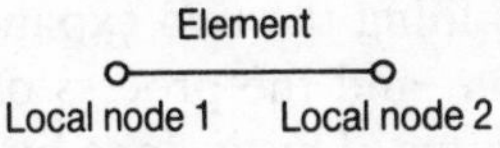

Element

○─────────────────○
Local node 1   Local node 2

$$\begin{pmatrix} a_{11} & a_{12} \\ a_{21} & a_{22} \end{pmatrix} \begin{pmatrix} u_1^{n+1} \\ u_2^{n+1} \end{pmatrix} = \begin{pmatrix} f_1 \\ f_2 \end{pmatrix}$$

(a) Single element

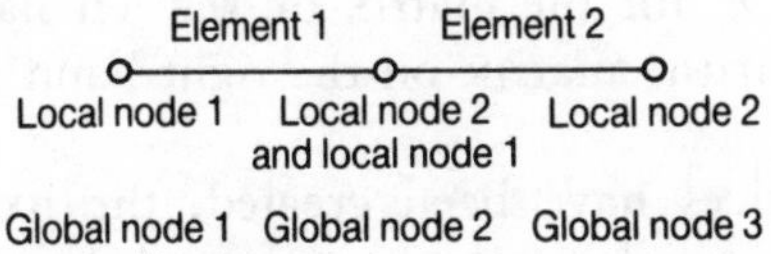

Element 1   Element 2

○─────────○─────────○
Local node 1   Local node 2   Local node 2
              and local node 1

Global node 1 Global node 2 Global node 3

$$\begin{pmatrix} a_{11} & a_{12} & 0 \\ a_{21} & a_{22} & 0 \\ 0 & 0 & 0 \end{pmatrix} \begin{pmatrix} u_1^{n+1} \\ u_2^{n+1} \\ u_3^{n+1} \end{pmatrix} = \begin{pmatrix} f_1 \\ f_2 \\ 0 \end{pmatrix}$$

$$\begin{pmatrix} 0 & 0 & 0 \\ 0 & b_{11} & b_{12} \\ 0 & b_{21} & b_{22} \end{pmatrix} \begin{pmatrix} u_1^{n+1} \\ u_2^{n+1} \\ u_3^{n+1} \end{pmatrix} = \begin{pmatrix} 0 \\ g_1 \\ g_2 \end{pmatrix}$$

giving

$$\begin{pmatrix} a_{11} & a_{12} & 0 \\ a_{21} & a_{22}+b_{11} & b_{12} \\ 0 & b_{21} & b_{22} \end{pmatrix} \begin{pmatrix} u_1^{n+1} \\ u_2^{n+1} \\ u_3^{n+1} \end{pmatrix} = \begin{pmatrix} f_1 \\ f_2+g_1 \\ g_2 \end{pmatrix}$$

(b)

Figure 3.6. Assembling element equations

This equation can be expressed as a matrix equation as shown in Figure 3.6(a), where the terms a_{ij} are functions of position derived from the integration of the first term on the left hand side of equation (3.24), and the terms f_i come from all the other terms in equation (3.24). This matrix equation is, in fact, part of a larger matrix equation for all the unknown values of U. Once all the equations for each element, the so-called *element equations*, are known then the full set of equations for the whole problem has to be produced. This is shown in Figure 3.6(b) where two elements are shown together with an expanded version of the element equations. These expanded equations are formed by relating the *local* node on an element to its *global* node number. For example, on element 2 the local node numbered

43

1 is global node number 2. Combining the two expanded equations produces a global matrix equation, and the process of combination is known as *assembling* the equations. This is done by adding all the element equations together as shown. The structural origins of the finite element method are apparent as the names of the matrices are taken from those that would be formed if a force acted on a set of springs. These names are, for the matrix on the left hand side, the *stiffness matrix* and, for the matrix on the right hand side, the *load vector*.

Once these global matrices have been created, the fixed-value boundary conditions are imposed on the matrices and the equations can be solved. Again, the solution of the original partial differential equation (3.16) has been reduced to the solution of a set of simultaneous equations. This may seem strange as the solution scheme is an explicit one and so should not require such a solution. For this case the left hand side of the global equations can be diagonalized using a technique known as *mass lumping* (Zienkiewicz and Taylor, 1989), and so the solution can then be found without solving the simultaneous equations.

3.2.3 *Using finite volumes*

Now that we have looked at the use of both finite differences and finite elements, we can turn our attention to the finite volume method. In practice this can be seen as a combination of the two other methods. As a first step in the transformation process, the forward difference in time is used to transform the left hand side of equation (3.16), just as we did with the finite element method. Then we form a finite volume in the x-direction. For simplicity, we will only look at the values at the nth time level. A typical finite volume, or cell, is shown in Figure 3.7, where the centroid of the volume, point P, is the reference point at which we wish to find a numerical analogue of the partial differential equation.

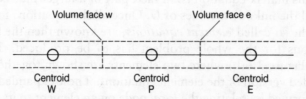

Figure 3.7. A finite volume in one dimension

Directions in the domain about the reference point are denoted by the points of the compass and so the neighbouring volumes are said to have their centroids at W and E, i.e. to the West and East of P if we consider the top of the diagram to be North. As the one-dimensional finite volume is centered on P, it will have one boundary face placed mid-way between the points W and P at the point labelled w, and another boundary face between P and E at the point e.

The spatial derivative is dealt with by noting that the second derivative of a variable at P can be taken as the difference of the first derivatives of the variable that are calculated at the volume faces, which gives

$$\left(\frac{\partial^2 U}{\partial x^2}\right)_P = \frac{\left(\frac{\partial U}{\partial x_e} - \frac{\partial U}{\partial x_w}\right)}{x_e - x_w} \tag{3.25}$$

Here, the subscripts refer to the positions at which quantities are either calculated or known. Similarly, the first derivatives at the volume faces can be taken to be the differences in the values of the variable at the neighbouring volume centroids, to give

$$\left(\frac{\partial U}{\partial x}\right)_e = \frac{u_E - u_P}{x_E - x_P} \tag{3.26}$$

and

$$\left(\frac{\partial U}{\partial x}\right)_w = \frac{u_P - u_W}{x_P - x_W} \tag{3.27}$$

Now that we have these three expressions for the various derivatives, they can be used to produce the numerical analogue of equation (3.16) at the point P. This analogue can be formed using any suitable version of the weighted average technique that we used with the finite difference transformation, giving either an explicit or an implicit scheme. Then the same techniques can be used to proceed once the initial and boundary conditions are known. When using finite volumes, all that is different is the philosophy behind the discretization procedure.

3.3 Comparison of the discretization techniques

From our short study of the application of these three numerical discretization methods to a simple partial differential equation, we can see that there are several common features. These features are that each method:

45

- produces equations for the values of the variable at a finite number of points in the domain under consideration;
- requires that we have a set of initial conditions to start the calculation for this time-dependent problem;
- requires that we know the boundary conditions of the problem so that we can find the values of the variables at the boundaries;
- can produce explicit or implicit forms and, if an implicit form is produced, then a set of simultaneous equations must be solved.

There are, however, several differences between the three methods and these include the following:

- The finite difference method and the finite volume method both produce the numerical equations at a given point based on the values at neighbouring points, whereas the finite element method produces equations for each element independently of all the other elements. It is only when the finite element equations are collected together and assembled into the global matrices that the interaction between elements is taken into account.
- The finite element method takes care of derivative boundary conditions when the element equations are formed, and then the fixed values of variables must be applied to the global matrices. This contrasts with the other two methods which can easily apply the fixed-value boundary conditions by inserting the values into the solution, but must modify the equations to take account of any derivative boundary conditions.

When looking at the simple example of a time-varying problem in one spatial dimension the domain in space has been extremely simple. Consequently, one problem that we have not addressed is how each of these discretization techniques is used to produce numerical equations for two- and three-dimensional spatial domains. Fortunately, our discussion of this simple example can shed some light on this.

Finite difference methods are based on the substitution of difference equations for the partial derivatives in partial differential equations. These difference equations link the values of variables at a set of points to the derivatives and so a grid of points is used throughout the spatial domain. In the example we have just discussed the grid was a line of points evenly spaced throughout the domain at various time levels. The difference formulae can be easily extended to cater for a spacing that is not even throughout the domain, and the partial differential equations can be transformed to cater for other coordinate systems that are not Cartesian. The finite difference method requires, however, that the grid of points is topologically regular. This means that the grid must look cuboid in a topological sense. This will be explained in greater detail when we discuss mesh and grid generation in Chapter 6.

If distributions of points with a regular topology are used, then the calculation procedure carried out by a computer program is likely to be extremely efficient and hence very fast. This is because the programmer can take advantage of the fact that the topology of the grid is always the same. The grid indexing system is extremely simple, say i, j, k in three dimensions, and is based on a set of local axes through the grid. Hence, when it is required to produce equations at some reference point, the program can determine the location of data at the neighbouring points simply from the maximum values of i, j, k. For example, if the grid is two-dimensional and has five points in the x-direction and ten points in the y-direction it will be as shown in Figure 3.8. There the grid is labelled both with the values of the indices i and j and the storage position of the variables in a one-dimensional array. For example, the value of the variable at position $i = 2$ and $j = 4$ will be stored in array location number 14. This assumes that the computer array stores points in the vertical direction first. From this it is easy to see that the neighbouring points to a reference point in the y-direction will be one array location

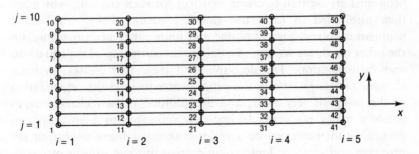

Numbers indicate array storage location

Figure 3.8. Data storage for rectangular grid

47

either forward or back from the reference position, and in the x-direction they will be ten points forward or back. An example of this is that from the value stored in array location 14 the value at the neighbouring point in the positive x-direction is stored at location 24. From this we see that simple arithmetic based on the topology of the grid is all that is required to find the location of the necessary values.

Finite elements, on the other hand, produce the numerical equations for each element from data at known points on the element and nowhere else. Consequently, there is no restriction on how the elements are connected so long as the faces of neighbouring elements are aligned correctly. By this we mean that the faces between elements should have the same nodes for each of the adjoining elements. This flexibility of element placement allows a group of elements to model very complex geometry, as we shall see later in Chapter 6.

Algorithms that have been developed using the finite volume method have tended to use a regular grid to take advantage of the efficiency of computation, just like the grids used with finite difference methods. Recently, however, to enable calculations to be carried out in complex geometries, algorithms have been developed with the finite volume method that can utilize irregular, finite-element-like meshes. It is the concept of the inter-volume flux across a face that enables this to be done. Both finite element and irregular-mesh finite volume programs pay a computational overhead for this geometrical flexibility, as look-up tables have to be used to find the geometrical relationships between the elements or volume faces, and this often involves finding data from the disk store of the computer. This overhead slows the programs down considerably.

One final advantage that the finite element method has is that the programs are written to create matrices for each element, which are then assembled to form the global equations before the whole problem is solved. Finite volume and finite difference programs, on the other hand, are written to combine the setting up of the equations and their solution. The decoupling of these two phases, in finite element programs, allows the programmer to keep the organization of the program very clear, and the addition of new element types is not a major problem. Adding new cell types to a finite volume program can, however, be a major task involving a rewrite of the program and so some finite volume programs can exhibit problems if they have multiple cell types.

3.4 Producing a solution from the discrete equations

Now that we have seen that discrete numerical equations can be formed from a partial differential equation using the three discretization methods that we have discussed, the next step is to solve these discrete equations to obtain a set of values for the variables at points in the domain. The ways that we use to do this must produce results that are both realistic and accurate. We talk of the methods converging and being stable. Also, if we use an implicit scheme, we must be able to solve sets of simultaneous equations. These subject areas are in the realm of the applied mathematician, and the discussion of them can be difficult to follow. However, the following texts do contain readable accounts of the techniques that are used, and these are Smith (1985), Zienkiewicz and Taylor (1989) and Hirsch (1988). The last of these three books contains much useful information about numerical discretization methods that is relevant to CFD.

When using CFD tools that have been written by someone else, we must hope that the software has been programmed to have a reliable means of producing a solution. However, CFD programs are so general that the user must intervene in the solution process and so some knowledge of the techniques that are used is necessary. In the following sections, some of the terminology and the techniques associated with the solution process are discussed.

3.4.1 *Convergence and stability*

Convergence and stability are two concepts that are often confused. In the strict mathematical sense convergence is the ability of a set of numerical equations to represent the analytical solution to a problem, if such a solution exists. The equations are said to converge if the numerical solution tends to the analytical solution as the grid spacing or element size reduces to zero. Equally, a process is stable if the equations move towards a converged solution such that any errors in the discrete solution do not swamp the results by growing as the numerical process proceeds.

In practice, however, these terms are used in less specific ways. For example, a numerical process is often said to converge if the values of the variables at the points in the domain tend to move towards some fixed value as the solution progresses. This use of the term 'convergence' arises because in most physical problems that we wish to solve with CFD there is no analytical solution to compare our numerical solution with. A process is said to be stable if this happens in such a way that the intermediate results of the process are reasonable. As was mentioned in Section 3.2.1 when we produced a numerical analogue of a partial differential equation using finite differences, the explicit solution scheme is only valid if the time step is sufficiently small. If the time step is too large, the values of the variables oscillate violently and become extremely large. This is an unstable process and it does not converge.

3.4.2 *Solving the simultaneous equations*

In most cases the discrete equations produced from partial differential equations are given in an implicit form. These implicit schemes are used because explicit schemes are less stable numerically, as we have discussed, and explicit schemes can produce results which diverge from physically realistic values as the solution progresses.

When implicit schemes are used a set of simultaneous equations is generated, consisting of many individual equations, and these must be solved in some way. There are many ways of doing this, and each software package will have its own way of producing a solution. In terms of computational effort the setting up of the equations might typically take half of the total computer time and the solution of the equations might take the other half. As the solving of the equations consumes a large amount of computational effort, there are great benefits to be gained from using fast methods of solving the simultaneous equations.

The solution of any set of simultaneous equations can be seen as the process of finding a vector x that satisfies the matrix equation

$$\mathbf{A}x = b \tag{3.28}$$

where $\mathbf{A}$ is an operator on the vector of variables x, and b is a vector of known values. The solution can be found by finding the inverse of the matrix $\mathbf{A}$ and then premultiplying both sides of equation (3.28) by the inverse. This gives

$$x = \mathbf{A}^{-1}b \tag{3.29}$$

If there are only a few equations in the set of simultaneous equations, then the inverse of the matrix $\mathbf{A}$ can be found easily and exactly. The methods used to do this are known as *direct methods* and, usually, they are versions of a method called LU decomposition as described by Zienkiewicz and Taylor (1989). In this method the matrix **A** is described by two other matrices in the following way

$$\mathbf{A} = \mathbf{LU} \tag{3.30}$$

where **L** is a lower triangular matrix and **U** an upper triangular matrix. Once the matrix **A** has been decomposed into **L** and **U** the solution is easy to find. If the matrix is large these direct methods require a lot of computer effort to produce a solution. This is the traditional way that finite element programs have produced their results. One way of reducing the computational effort is to use *iterative methods* of solution for large systems of equations. These take some guess for the values of the solution vector x and then produce a more accurate guess from the vector x and the coefficients of the matrix **A** and vector b.

A variety of iterative schemes are commonly used and some of these are discussed by Smith (1985) and Hirsch (1988). It helps when considering the solution of equation systems to think of a simple case. For example, if equation (3.28) is a system of three equations it could be rewritten as:

$$\left. \begin{array}{l} a_{11}x_1 + a_{12}x_2 + a_{13}x_3 = b_1 \\ a_{21}x_1 + a_{22}x_2 + a_{23}x_3 = b_2 \\ a_{31}x_1 + a_{32}x_2 + a_{33}x_3 = b_3 \end{array} \right\} \tag{3.31}$$

if the individual equations are listed separately. Using this we can start to identify some of the common iterative schemes such as the following:

- *Jacobi and Gauss–Seidel methods*. In these two methods the equations are rewritten as

$$\left. \begin{array}{l} x_1 = \dfrac{1}{a_{11}} \left(b_1 - a_{12}x_2 - a_{13}x_3 \right) \\[2ex] x_2 = \dfrac{1}{a_{22}} \left(b_2 - a_{21}x_1 - a_{23}x_3 \right) \\[2ex] x_3 = \dfrac{1}{a_{33}} \left(b_3 - a_{31}x_1 - a_{32}x_2 \right) \end{array} \right\} \tag{3.32}$$

from which we can see that the diagonal terms of matrix **A**, i.e. the terms a_{ii}, cannot be zero if these methods are to work. The Jacobi method takes the right hand side of equation (3.32) to be the known values at the nth iteration and the left hand side to be the new values at the $n + 1$th iteration, giving

$$x_1^{n+1} = \frac{1}{a_{11}} (b_1 - a_{12}x_2^n - a_{13}x_3^n)$$

$$x_2^{n+1} = \frac{1}{a_{22}} (b_2 - a_{21}x_1^n - a_{23}x_3^n) \qquad\qquad (3.33)$$

$$x_3^{n+1} = \frac{1}{a_{33}} (b_3 - a_{31}x_1^n - a_{32}x_2^n)$$

and the Gauss–Seidel method takes advantage of the fact that once a new value is known at the $n + 1$th iteration it can be used on the right hand side of the equations giving

$$x_1^{n+1} = \frac{1}{a_{11}} (b_1 - a_{12}x_2^n - a_{13}x_3^n)$$

$$x_2^{n+1} = \frac{1}{a_{22}} (b_2 - a_{21}x_1^{n+1} - a_{23}x_3^n) \qquad\qquad (3.34)$$

$$x_3^{n+1} = \frac{1}{a_{33}} (b_3 - a_{31}x_1^{n+1} - a_{32}x_2^{n+1})$$

Both these methods require that an initial guess to the solution is made which can then be used during the first iteration.

- *Point relaxation methods.* At any stage in the iteration procedure there will be a finite error in the solution vector x. One way of classifying this error is to use equation (3.28) to find what is known as the *residual error*, which is defined as

$$r = b - \mathbf{A}x \qquad\qquad (3.35)$$

This residual should become ever smaller as the iterations proceed and it can also be used in the iteration procedure. To do this we take the equations of the Gauss–Seidel method (equation (3.34)) and both add and then subtract the terms x_i^n to the right hand side. This gives

$$x_1^{n+1} = x_1^n + \left[\frac{1}{a_{11}} (b_1 - a_{11}x_1^n - a_{12}x_2^n - a_{13}x_3^n) \right]$$

$$x_2^{n+1} = x_2^n + \left[\frac{1}{a_{22}} (b_2 - a_{21}x_1^{n+1} - a_{22}x_2^n - a_{23}x_3^n) \right] \qquad (3.36)$$

$$x_3^{n+1} = x_3^n + \left[\frac{1}{a_{33}} (b_3 - a_{31}x_1^{n+1} - a_{32}x_2^{n+1} - a_{33}x_3^n) \right]$$

In these equations the expressions in square brackets are the terms of the residual r. As we know that these should tend

52

to zero as the iteration progresses there is no reason why we should not try and accelerate the process by multiplying the residual by some factor ω, which is known as a *relaxation factor*. This gives

$$\left.\begin{aligned} x_1^{n+1} &= x_1^n + \left[\frac{\omega}{a_{11}}(b_1 - a_{11}x_1^n - a_{12}x_2^n - a_{13}x_3^n)\right] \\ x_2^{n+1} &= x_2^n + \left[\frac{\omega}{a_{22}}(b_2 - a_{21}x_1^{n+1} - a_{22}x_2^n - a_{23}x_3^n)\right] \\ x_3^{n+1} &= x_3^n + \left[\frac{\omega}{a_{33}}(b_3 - a_{31}x_1^{n+1} - a_{32}x_2^{n+1} - a_{33}x_3^n)\right] \end{aligned}\right\} \quad (3.37)$$

and for most systems of equations the value of ω can be set to somewhere between the values of one and two. Hence the method is known as a *successive overrelaxation method*. If ω is unity the method becomes the original Gauss–Seidel method.

- *Line relaxation methods*. The methods given above generate a new estimate for the solution vector x one term at a time, which is very similar to the explicit methods we have already discussed. Sometimes it is possible to speed up the process if a small sub-set of the terms is found simultaneously. This is an implicit way of proceeding and involves the direct solution of a smaller set of equations. The commonest way of doing this is to take the solution at a whole line of points in a regular grid describing a spatial domain, and solve line by line rather than point by point. Equally, if a regular three-dimensional grid is used, a rectangular slab of points could be calculated directly in one step of the iteration process.

- *More advanced methods*. As further research into the iterative solution of simultaneous equations takes place more methods of solution emerge. This is driven by the need to reduce the computational effort required to solve the large systems of equations on supercomputers where the effort is still excessive for many engineering problems. These advanced methods include Stone's strongly implicit procedure and preconditioning methods, which can be seen as matrix manipulation procedures, and multigrid methods which calculate the solution on a series of coarse and fine grids in space, swapping between the grids in such a way that any errors are smoothed out.

As users of CFD software our concern with the solution of the simultaneous equations that are generated will usually be restricted to providing some of the controlling parameters for the solution methods built into the software. It should be noted here that if the solution method is an iterative one the exact values of the vector x may never be found, but that after a few iterations the error in x should be very small. Also, as we shall discuss later, the fluid flow equations are non-linear and possibly time-dependent, and so we will require the solution procedure to find successive approximations to the flow variables regardless of whether we solve the equations themselves in a direct or iterative way. This means that the solution to the simultaneous equations generated need only be approximate, giving some improvement in the values of the variables.

3.5 Solving the coupled set of fluid flow equations

In this chapter we have considered the discretization of general partial differential equations and the solution of the numerical analogue. Now it is time to look at the numerical solution of the partial differential equations that govern fluid flow. These equations were presented in Chapter 2 and they can be discretized using any of the three discretization techniques that we have already discussed. The numerical analogues of the original partial differential equations have then to be solved. For reasons that we will now discuss, the equations governing fluid flow are particularly difficult to discretize and solve using numerical techniques.

3.5.1 *Non-linearity and time dependence*

For a two-dimensional flow problem we have to solve two momentum equations and the continuity equation. That is, we have three equations which we can use to find the three flow variables which are the velocity components u and v of the fluid and the fluid pressure p. The two momentum equations are time-dependent and they are also non-linear. The non-linearity comes from the convection terms for the velocity components that are derived from the acceleration of a patch of fluid. These two factors of time dependence and non-linearity increase the complexity of the solution.

Dealing with time dependence is handled in the same way that it was handled for the simple parabolic partial differential equation discussed earlier in Section 3.2. We must know the initial conditions of the problem to enable our solution to begin, and from these the solution at the next time level is found. This means that our solution procedure proceeds via a series of iterations in time.

At each time step the equations are non-linear and so we must linearize them so that a solution can be found to a set of simultaneous equations which look like the form we have just discussed, i.e.

$$\mathbf{A}x = b \tag{3.38}$$

but where the matrix $\mathbf{A}$ and the vector b are functions of the flow variables. The linearization is carried out by discretizing the derivative that appears in the convection terms as normal and taking the current value of velocity at a point or in a volume or element as the velocity multiplier. For example,

$$u \frac{\partial u}{\partial x} \qquad \text{becomes} \qquad \bar{u}\left(\frac{u_{i+1,j} - u_{i-1,j}}{2\delta x}\right) \tag{3.39}$$

if the central difference equation (3.4) is used for the derivative and $\bar{u}$ is found from the current solution for U. For example, $\bar{u}$ would be $u_{i,j}$ if we were using a finite difference method. Once this linearization is carried out the set of simultaneous equations can be produced and then solved to update the values of the flow variables. The linearization and solution procedure is then repeated until the values of the flow variables have converged, and only then can the whole solution be progressed to the next time level.

From this we can see that there are several levels of iterative process taking place within the solution algorithm. Figure 3.9 shows these levels schematically. There we can see that there is an outer time iteration loop that moves the solution from one time level to the next. Then there is an inner loop that resolves the non-linearity in the equations by repeatedly forming sets of linear simultaneous equations. This loop might itself contain a further loop where iterative methods are used to solve the simultaneous equations that are generated.

When running steady state fluid flow examples the time iteration loop can be left out of the process. However, the absence of the time terms in the momentum equations can cause numerical problems as the fluid acceleration is not modelled in the same way as it would occur for a physical flow. This can lead to a common problem where the numerical solution will not be stable and so it will diverge from reality. As the non-linearity of the problem forces us to use an iterative solution scheme, there is no real advantage to be gained by leaving the time terms out. Consequently, many CFD programs use

55

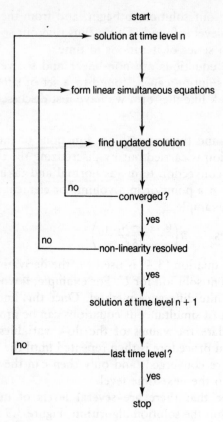

Figure 3.9. An iterative solution

a time-dependent algorithm even for steady state cases and this enhances the stability of the method.

3.5.2 *Obtaining the pressure solution*

Having looked at the overall solution process that must take place to solve the governing equations, we must now look in more detail at how to obtain the solution. If we look at the three equations that govern two-dimensional incompressible fluid flow, we can see that the two momentum equations contain all three flow variables, but that the continuity equation contains only the velocity components. As most of the terms in the momentum equations are functions of the velocity components it is natural to use these equations to produce the solutions for the velocity components. This then leaves a problem in that the continuity equation does not contain terms that include the fluid pressure.

One way of overcoming this problem is to discretize the three equations in such a way that they can be solved together. This leads to a solution vector that contains all three variables and so is three times longer than it need be, but it does allow the pressure to be calculated. Finite element programs have been developed in this way for some time, but as this approach produces larger matrices than would be generated if each variable was solved for in turn, a larger amount of computer effort is required to produce the solution.

An alternative approach is to discretize the momentum equation in the x-direction so that the u-velocity component can be found, and similarly find the v-velocity component from the momentum equation in the y-direction. Then a modified form of the continuity equation has to be developed so that the pressure can be calculated. This is done by noting that the velocity components that are found from the momentum equations do not satisfy the continuity equation and that they should satisfy this equation when the solution is converged. If the variables are split into two parts, the values that satisfy the momentum equations (starred) and the corrections that would ensure that continuity is satisfied (dashed), we can write:

$$\left.\begin{array}{l} u = u^* + u' \\ v = v^* + v' \\ p = p^* + p' \end{array}\right\} \tag{3.40}$$

As during the solution procedure, we have to ensure that the continuity equation, equation (2.2), is satisfied; we can take that equation

$$\frac{\partial u}{\partial x} + \frac{\partial v}{\partial y} = 0 \tag{3.41}$$

and then substitute into it the expressions in equation (3.40) to give

$$\frac{\partial u'}{\partial x} + \frac{\partial v'}{\partial y} = -\frac{\partial u^*}{\partial x} - \frac{\partial v^*}{\partial y} \tag{3.42}$$

In this equation the derivatives of the correction velocity components depend on the derivatives of the velocity components that satisfy the momentum equations. Now when the momentum equations (2.8) and (2.9) are discretized they can also be written in matrix form as

$$\mathbf{A}u_j = \mathbf{B}p_j \tag{3.43}$$

57

and

$$Cv_j = Dp_j \qquad (3.44)$$

where **A**, **B**, **C** and **D** are matrices, and u_j, v_j and p_j are vectors of the variables at grid points or nodes. These equations can be rewritten if the variables are split using equation (3.40), to give

$$Au_j^* + Au_j' = Bp_j^* + Bp_j' \qquad (3.45)$$

and

$$Cv_j^* + Cv_j' = Dp_j^* + Dp_j' \qquad (3.46)$$

When we solve the momentum equations we are in effect solving the following two equations:

$$Au_j^* = Bp_j^* \qquad (3.47)$$

and

$$Cv_j^* = Dp_j^* \qquad (3.48)$$

and so these can be subtracted from the matrix equations (3.45) and (3.46), giving

$$Au_j' = Bp_j' \qquad (3.49)$$

and

$$Cv_j' = Dp_j' \qquad (3.50)$$

It is these two equations that are the expressions that enable the correction quantities for the velocity components to be found, as they can be rewritten to give

$$u_j' = A^{-1}Bp_j' \qquad (3.51)$$

and

$$v_j' = C^{-1}Dp_j' \qquad (3.52)$$

Using these two forms of the equations we can find the pressure from the continuity equation. This is done by substituting them into the modified continuity equation (3.42), to produce an equation for the correction pressure p_j' which has on its right hand side the imbalance in the continuity of the flow after the momentum equations have been solved. Once the correction pressure p' has been found, so u' and v' can be found using equations (3.51) and (3.52). Finally equations (3.40) are used to find the corrected velocity components and pressure. At this stage in the solution the velocity components satisfy the continuity equation and a new value of pressure has been calculated, but the velocity components do not satisfy the momentum equations. To resolve both the solution of the momentum equations and the non-linearity, the momentum equations are used again to produce further simultaneous equations which are solved, followed by the calculation of the correction pressure and the correction velocities. It is this process of using the momentum equations then the continuity equation that forms the inner loop in Figure 3.9, and iterative methods are used to solve all three sets of simultaneous equations within each inner iteration.

Algorithms such as this are known as SIMPLE (Semi-Implicit Pressure Linked Equations) algorithms and there are many variants of the algorithm described above where small modifications are made to the procedure.

Having found a way of obtaining the pressure solution, there is only one remaining problem to solve. This concerns the numerical solution of the equations. Looking at the momentum equations (2.8) and (2.9) we can see that the pressure variable only occurs in a first-order spatial derivative. The conversion of these derivatives to numerical form can lead to problems, as the use of central differences can produce values for the pressure variable at a given point which are not related to the pressure variables at neighbouring points. This, in turn, can lead to a pressure solution oscillating in what is known as a *chequerboard* pattern. There are ways of overcoming this and many programs use a grid which is staggered from the grid for the velocity components to find the pressure. Effectively, the pressure is stored at the centroid of a volume and the velocity components are stored at the volume faces (Patankar, 1980). More recently, several programs have turned to storing all the variables at volume centroids using the transformation of Rhie and Chow (1983) to prevent chequerboarding.

3.5.3 *The convection operator*

One other problem that has had to be addressed by researchers is that of producing numerical forms of the convection operator. Problems occur when this operator is discretized using central differences, equation (3.4), for the first derivative of the velocity. For example, take the equation

$$\bar{u}\,\frac{\partial u}{\partial x} = v\,\frac{\partial^2 u}{\partial x^2} \tag{3.53}$$

where $\bar{u}$ denotes the known velocity that is being used to linearize the equation. Using central differences for both the first and second derivatives in this equation gives

$$\bar{u}\left(\frac{u_{i+1,j} - u_{i-1,j}}{2\delta x}\right) = v\left(\frac{u_{i+1,j} - 2u_{i,j} + u_{i-1,j}}{\delta x^2}\right) \tag{3.54}$$

which can be rearranged to give

$$u_{i,j} = \tfrac{1}{2}u_{i+1,j}\left(1 - \frac{Pe}{2}\right) + \tfrac{1}{2}u_{i-1,j}\left(1 + \frac{Pe}{2}\right) \tag{3.55}$$

where *Pe* is the Peclet number, or local cell Reynolds number, given by

$$Pe = \frac{\bar{u}\delta x}{v} \tag{3.56}$$

From equation (3.55) we can see that the value of the Peclet number has an important effect on the numerical equation. When the Peclet number is less than two, both terms on the right hand side have positive coefficients, but when the Peclet number is greater than two, the first term on the right hand side becomes negative. This negative term causes problems in that it can lead to unrealistic solutions. Consequently, there is a restriction on the Peclet number if we want to get realistic values.

One way around this is to use a first-order accurate difference equation to model the first derivative in equation (3.53) instead of the second-order accurate difference equation used above. However, the reduction in accuracy can lead to a poor solution. Typically the use of lower-order accuracy schemes gives results which are the results for a flow which has more viscosity than the one we are trying to model. Despite this such schemes are in common use together with more accurate schemes. Usually commercial CFD packages will have one of the following options for the discretization of the convection operator:

- *An upwind scheme*, where the convection term is formed using a first-order accurate difference equation equating the velocity derivative to the values at the reference point and its nearest neighbour taken in the upstream direction. This can give very inaccurate solutions but they are easy to obtain as they converge readily.
- *A hybrid scheme*, where the upwind scheme is used if the Peclet number is greater than two, and central differences are used if the Peclet number is two or less. This is more accurate than the upwind scheme but does not converge on some grids of points.

- *QUICK*, which is a quadratic upwind scheme and is more accurate than the two schemes described above. For complex geometries the shape of the volumes can lead to numerical problems in obtaining the solution.
- *Power-law schemes*, which are derivatives of QUICK but are more accurate.

A good review of this topic is given by Abbott and Basco (1989).

3.5.4 *Boundary conditions for fluid flow problems*

When solving any system of partial differential equations it is the boundary conditions, together with the initial conditions, that determine the exact solution. The form of the boundary conditions that is required by any partial differential equation depends on the equation itself and the way that it has been discretized. Some common boundary conditions are, however, met when solving fluid flow problems with computers. These can be classified either in terms of the numerical values that have to be set or in terms of the physical type of the boundary condition.

Looking at the variables, we need boundary conditions for the following variables:

- For the velocity components, which will affect the momentum equations. These conditions are usually given by specifying the velocity components and if this is not done then the derivatives of the velocity components normal to the boundary are usually zero.
- For the pressure and, possibly, mass flow, which will influence the continuity equation if a SIMPLE-like algorithm is being used. Usually, the fluid pressure needs to be specified at a minimum of one point in the flow.
- For the turbulence variables such as the turbulence kinetic energy k and the rate of dissipation of k, i.e. ε.

These conditions have to be applied at a variety of boundaries such as the following:

- *Solid walls*. Many boundaries within a fluid flow domain will be solid walls, and these can be either stationary or moving walls. If the flow is laminar then the velocity components can be set to be the velocity of the wall. When the flow is turbulent, however, the situation is more complex. This

61

complexity is due to the velocity of a flow varying extremely rapidly near a wall if the flow is turbulent. To capture this rapid variation, which occurs in a direction away from the wall, many grid points would be required in this direction near the wall, and this increases the amount of computational effort required to produce a solution. One way of reducing the effort is to specify the velocity near a solid wall using experimental data for boundary layers which shows that the velocity variation should be logarithmic with the distance from the wall at points more than a known distance from the wall. This can be seen in Figure 3.10 where the velocity in the boundary layer is plotted against distance away from the wall. Both the velocity and distance have been transformed into non-dimensional quantities as shown. Looking at the diagram, three regions can be seen. Near the wall there is a *viscous sub-layer* where the effects of turbulence are damped out by the wall itself. Then there is a *log-law region* where the velocity is a logarithmic function of the distance from the wall. Finally, there is an *outer layer* which is where the boundary layer and the external flow merge. If the mesh is built so that the first point where the velocity is calculated is in the log-law region, then the very rapid variation near the wall will not need to be modelled. Similar methods can be used to specify the values of both the turbulence variables k and ε.

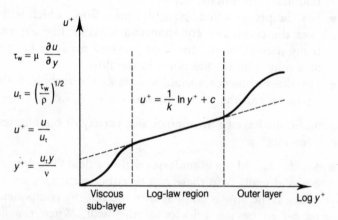

$$\tau_w = \mu \frac{\partial u}{\partial y}$$

$$u_\tau = \left(\frac{\tau_w}{\rho}\right)^{1/2}$$

$$u^+ = \frac{u}{u_\tau}$$

$$y^+ = \frac{u_\tau y}{\nu}$$

$$u^+ = \frac{1}{k}\ln y^+ + c$$

Viscous sub-layer   Log-law region   Outer layer   Log y^+

Figure 3.10. Velocity variation near a wall

62

Solving the coupled set of fluid flow equations

- *Inlets*. At an inlet, fluid enters the domain and so the fluid velocity might well be known for the problem being simulated. In some programs the pressure equation can only be solved if the mass flow at an inlet is known. Also, the fluid carries with it other quantities such as k and ε and so these must be specified as well. We say that variables are convected into the domain.

- *Outlets*. Where the fluid leaves the domain is known as an outlet. Normally, the pressure is set to zero at an outlet and the velocity components and any turbulence variables are left to find their own values which will have a zero spatial derivative in a direction normal to the boundary. If the flow is swirling through the outlet then a pressure gradient is required to provide the necessary centripetal force to the fluid and so a constant pressure boundary condition will be

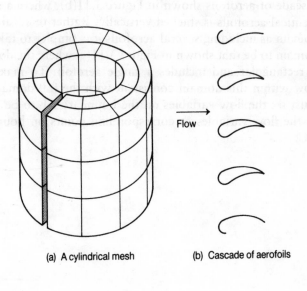

Flow

(a) A cylindrical mesh   (b) Cascade of aerofoils

(c) Domain around a single aerofoil

Figure 3.11. Cyclic boundaries

63

invalid. To overcome this, iterative procedures are used which start by specifying a constant pressure at the outlet but then try to find the pressure that matches the velocity of the swirling flow.

- *Symmetry boundaries.* When the flow is symmetrical about some plane there is no flow through the boundary and the derivatives of the variables normal to the boundary are zero.
- *Cyclic or periodic boundaries.* These boundaries come in pairs and are used to specify that the flow has the same values of the variables at equivalent positions on both of the boundaries. In Figure 3.11 two examples of periodic boundaries are shown. In the first (Figure 3.11(a)) a mesh which is topologically cuboid has been wrapped around onto itself. On the shaded boundary and the boundary facing it the fluid variables must be set to be equal at the corresponding points. The other example concerns the cascade of aerofoils shown in Figure 3.11(b), where a set of identical aerofoils is stacked vertically. Rather than take the domain as including several aerofoils it is simpler to take the domain to be that shown in Figure 3.11(c) where the domain is rectangular and includes a single aerofoil. To make the flow within this domain consistent with the full domain we must set the flow variables on the boundary AB to be equal to the flow variables at corresponding points on boundary DC.

4 Computer-based analysis procedures and tools

In the last two chapters we have looked at the ways in which fluid dynamics problems can be described mathematically and at how the governing equations can be transformed to give a numerical analogue. To produce a solution from the numerical analogue many equations have to be calculated and this in turn requires vast numbers of repetitive calculations to be carried out. Computers are the ideal tool for this numerical processing as they can be programmed to perform all of these calculations without intervention. We saw in Chapter 3 that the numerical solution process is specific to the equations that are solved and not to the actual flow problem being simulated. It is the boundary conditions and the initial conditions that are applied that determine the flow problem.

Within a given class of flow problems, say, for example, those that have a flow which can be taken to be viscous and incompressible, general computer software can be written to produce solutions to the governing equations and this software is not problem-specific. Many industrial organizations require information on flow situations and so they either write their own CFD simulation program or buy one of the software packages written by a specialist software company. As there is a growing commercial market for these programs there are several available.

Not only can the software be general to a flow type, but also the analysis process that is followed can be general too. This means that regardless of the software being used there is a clearly defined set of stages that make up the analysis process. The first section of this chapter defines this process by looking at the material that we have already discussed in the last two chapters and then determining what the key stages of the CFD analysis process are. Here only an overview of the process is given but in subsequent chapters we will discuss each stage of the process in detail. This is followed by a series of examples that show the process at work.

As the analysis process centres on computational procedures, the analyst has to use a wide variety of hardware and software tools and, again, these can be classified into sets of standard types. So the second section of this chapter looks at the types of hardware installation that can be used to run a CFD problem. This is followed by a section on the use of the hardware and then the final section discusses the CFD software tools that are available. In all of this there is a recurring theme of the universality of the analysis process and the tools that are used in carrying it out.

4.1 The analysis process

We have seen in Chapter 2 that a mathematical analysis of fluid flow can be made and that this leads to a series of partial differential equations that govern the flow. In Chapter 3 we saw that these partial differential equations can be discretized to produce a numerical analogue of the equations. When boundary conditions and initial conditions that are specific to the flow problem being simulated have been applied to these equations, they can be solved using a variety of direct or iterative solution techniques producing a numerical simulation of the given flow problem. Many of the numerical aspects of the flow simulation are handled by the CFD computer programs that have been written, but the user of these programs must specify several pieces of information to the program in order that a successful simulation can be made.

From our study of the ways in which it is possible to produce numerical solutions to the governing partial differential equations we have found that the following are required if a solution is to be produced:

- a grid of points, or a set of volumes or elements, at which to store the variables that need to be calculated,
- boundary conditions that enable the boundary values of the variables to be calculated,
- initial conditions that define the initial state of the flow for a transient problem or define the first guess to the variables for a steady state problem,
- fluid properties that appear in the equations such as density and viscosity and perhaps some turbulence quantities,

- control parameters that affect the numerical solution of the equations,

and it is the provision of this information that dictates the stages of the analysis process. Given that the analyst has the necessary hardware and software tools, the stages of the analysis process that must be carried out to generate this information and then produce the results of the flow simulation are as follows:

- *Initial thinking.* As with many analytical or computational problems it is worth thinking about the physics of the problem for a while before committing pen to paper or fingers to keyboard. In this first stage the analyst should consider the flow problem and try to understand as much as possible about it. This might involve a considerable amount of liaison with any other people involved in the project such as design engineers and technicians, but it is important that all sources of relevant information are explored.

- *Mesh generation.* In this stage the analyst has to calculate the grid of points or mesh that sub-divides the flow domain. A series of coordinates for the points in the mesh have to be calculated and sometimes these points must be related to define the volumes, also known as cells, and elements. It is the distribution of the points in the flow domain that defines the positions where the flow variables are calculated.

- *Flow specification.* Once a mesh exists, the boundaries of the computational domain can be found and the necessary boundary conditions, determined in the initial phase, applied. These conditions, together with the initial conditions and some fluid parameters, specify the actual flow problem that is to be solved.

- *Calculation of the numerical solution.* Now the CFD software can be run to calculate the numerical solution to the flow problem, but first the user must provide the information that will control the numerical solution.

- *Results analysis.* When some results have been obtained they must be analysed, first to check that the solution is satisfactory and then to determine the actual flow data that is required from the simulation.

It is possible to perform an analysis by taking each stage in the order given above, so that the required results are generated. This would only happen in an ideal situation as the simulation of flow problems can be extremely difficult. The governing equations are complex, as they are non-linear and highly coupled, and can be time-dependent. This means that the possibilities for some error creeping into the solution procedure are great, leading to a simulation that will not converge or to a set of results that are not very good. These problems can be reduced by a combination of user experience and good practice during the analysis. By 'good practice' we mean that the analysis should be carried out extremely carefully so that the analyst makes sure that each stage is completed successfully before proceeding to the next stage.

Working in this way will usually mean that the analyst makes a series of checks during each stage. The necessary checks will be described in the subsequent chapters, but if they show that the simulation is not progressing well then it may be necessary to repeat one or more of the stages. By doing this the computer model can be modified in an attempt to improve the simulation. As computers are

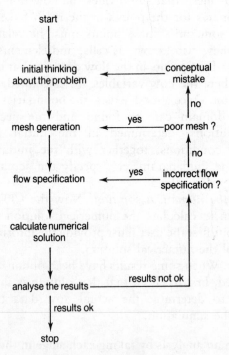

Figure 4.1. The analysis process

being used the refinement of the computer model is not too difficult to perform, as, usually, information stored on computers can be accessed and changed both quickly and easily. This interaction between the stages of the analysis process should enable reliable results to be achieved, given the constraints of the hardware and software. Some of these interactions are shown in the flowchart of the process given in Figure 4.1. There the importance of careful analysis of the results can be seen. By looking at the results produced it is possible to see if a simulation is a good one. If it is not then the flow specification might be incorrect, or the mesh might not be suitable for the flow being modelled, or a conceptual mistake could have been made at the beginning of the analysis process.

4.2 Computer hardware for CFD

4.2.1 *Computers*

As we have seen, we need computers to perform the repetitive calculations that produce the solution to the numerical equations. As computer technology changes at an alarming rate the supercomputers of one era become the desktop calculators of the next and, consequently, we need to be wary of reviewing the state of the art in computer technology. Even so, we can still produce a series of classifications that describe the generic hardware types that are available.

In the world of engineering computation it is common to classify computers by their performance in terms of some measure of calculation speed. Speed can be measured in units based on the number of instructions that a processor can execute per second or the number of floating point operations that a system can handle per second. Common units are *mips* or millions of instructions per second and *MFLOPS* or millions of floating point operations per second. These measures can give a user some idea of the throughput of a machine but they say nothing about the ways in which the systems operate with a particular numerical software package. This is very important when we consider the CFD analysis process, as the execution of calculations is only one part of the process of producing a final solution. Other features such as the speed of access of data are equally important to the overall speed of the calculation.

If we consider the operational characteristics of the computers that are used to perform CFD calculations we can divide the computer

69

types into the following five categories:

- *Personal computers.* These are stand-alone systems containing a central processor, some random access memory (RAM) and some disk storage. Usually they have single-user operating systems.
- *Workstations.* These are machines that have a central processor, local RAM storage, and multi-user operating systems. These are packaged together with a high resolution graphics display. They are often part of a network of machines that can include a central data storage system such as several disks attached to a *file server*, which is a computer that is dedicated to the task of providing datafiles to the other machines in the network. The network can also be used to gain access to high speed computers and a variety of peripherals. Some workstations also have their own local disk storage.
- *Mini-computers.* These are machines with a central processor, large amounts of RAM storage and a central data storage system. They have multi-user operating systems and are used by several people simultaneously who gain access to the system by using terminals.
- *Mini-supercomputers.* These are effectively super-workstations, with very good graphics performance and near-supercomputer numerical performance. Again, they are usually part of a network.
- *Supercomputers.* Designed to handle numerical data in the fastest possible way, these machines are dedicated to the task of running numerical simulations. They are large, high technology devices often with multiple processors and extremely large amounts of RAM storage to reduce the need for the machine to communicate with slower storage devices when carrying out calculations. To enable good graphics facilities to be used, supercomputers are often networked to workstations.

Figure 4.2 shows the configurations of the machine types in this list, but this is clearly not a full list as other types of machine are available. In particular, machines that have a large number of processors such as transputers have a great potential to carry out numerical calculations extremely quickly. At the time of writing (1991) the CFD calculations that are carried out on these machines

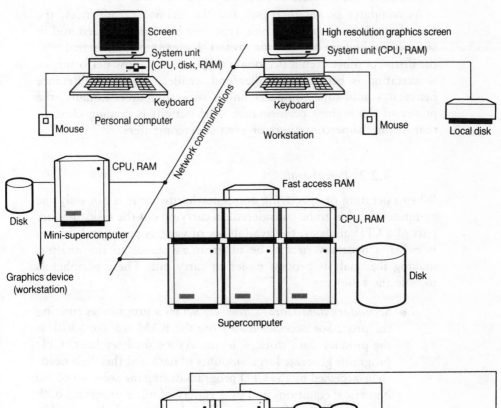

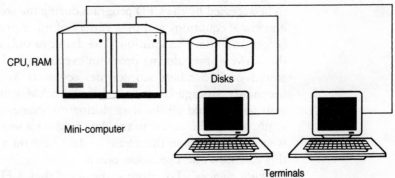

Figure 4.2. Computer types

use specialist programs adapted to take advantage of the internal architecture of the hardware. Commercial programs are rarely available on these machines, but no doubt this will change.

As computer power increases and the technology advances, the boundaries between the machine types are becoming blurred and in some cases they are being made invalid. For example, again in 1991, the distinction between a personal computer and a low performance workstation is becoming smaller and smaller, as is the difference between a mini-supercomputer and a supercomputer. Equally, the power of the highest performance workstations is coming close to that of mini-supercomputers or even supercomputers.

4.2.2 *Peripherals*

When operating or specifying computer hardware it is not only the computer that has to be considered. In carrying out the tasks that are part of a CFD analysis, the availability of various peripheral devices is either a necessity or can be of great assistance to the analyst, making the analysis process easier to carry out. These peripherals include the following:

- *Secondary data storage devices.* When a program is running the processor accesses data from the RAM storage which is the primary data storage device. As we shall see later, CFD programs generate large amounts of data and this data needs to be accessed by the CFD program during the solution of the numerical equations and by a variety of other programs both before and after the solution. The data can only be stored in the RAM storage during program execution and so it is also stored on secondary storage devices such as hard disks. Secondary storage is also used if the RAM capacity is not sufficient to hold all the data during the execution of a program, and as the access to the data held on a secondary store is much slower than the access to data held on a RAM store this can slow the execution down.
- *Backup devices.* To protect the data that CFD programs generate from loss due to a failure of a disk drive or a disaster such as a machine-room fire, it is necessary to make regular copies of the data onto some form of backup data store, for example demountable hard disks, magnetic tapes or other devices which can be removed to a safe storage area. Often,

this is done automatically, or is handled by the administrator of the computing system.

- *High resolution graphics displays.* CFD analyses generate so much data that, quite often, the only way of analysing the data over the whole domain is to use some form of graphical representation. High resolution graphics display devices are used to show the necessary pictures and these devices include the screen of a workstation or a dedicated graphics terminal. Typically, the resolution of these devices is 1000×1000 *pixels*, where a pixel is a dot on the screen, although useful work can be done at lower resolutions such as 600×400 pixels. The addition of colour can also be extremely helpful in clarifying the pictures produced from large CFD-generated databases.

- *Hardcopy devices.* As with most engineering activities, report writing is a necessary evil and so a means of obtaining a hardcopy of the pictures generated on a graphics device is necessary. These copies can come from a laser printer which will produce black and white copies or from a colour plotter which uses ink-jets or heated waxes to produce a coloured image.

4.3 Using the hardware

When we simulate fluid flow problems with a computer the analysis process has three main requirements that have an impact on the computer hardware. We saw in Chapter 3 that the solution process consists of calculations that are carried out at a large number of points in the domain under consideration and that these calculations are part of an iterative process in which individual calculations have to be repeated many times before the solution is obtained. Consequently, the numerical calculation phase of the simulation requires both a large amount of data storage and considerable computer processing power if real-life engineering problems are to be solved. Further, the large amount of data that is generated has to be analysed graphically and so the computer system must support the production of graphical data as well.

Now, by referring to these requirements of data storage, processing power and graphics capabilities, each of the computer classifications that we listed earlier can be analysed in turn. As can be seen from the following:

- Personal computers do not have the processing power or data storage capacity to enable large simulations to be run. These machines might be used for training problems, where the size of the problem is very small and the speed of analysis is unimportant.
- Workstations have all the necessary computing power, data storage and graphics capabilities for some problems. However, the largest problems may require extra computing power such as that provided by a mini-supercomputer or a supercomputer.
- Mini-computers tend to perform like workstations but they sometimes have less graphics capability.
- Mini-supercomputers can be used for most problems, including the largest simulations, if the turnaround time of the numerical analysis is not too important.
- Supercomputers are especially useful for the largest problems where results are required quickly. They tend to be linked to workstations, or mini-computers, to enable the graphics tasks of the analysis to be carried out on a smaller machine, leaving the raw computing power and large data storage for the numerical applications that need them.

This situation is summarized in Table 4.1. It should be emphasized that useful CFD analyses can be carried out with a limited amount of hardware and so it is not necessary to have access to a supercomputer to start using CFD. Situations do exist, however, where access to a supercomputer might be required if simulations are to be

Table 4.1. *Suitability of hardware types*

Hardware type	Graphics	Small jobs	Medium jobs	Large jobs
Personal computer	OK	OK	No	No
Workstation	Good	Good	Good	No
Mini-computer	OK	Good	Good	No
Mini-supercomputer	Good	Good	Good	OK
Supercomputer	No	OK (expensive!)	Good	Good

achieved in a reasonable time scale. Guidelines for the specification of hardware are given in Chapter 12, but it is worth noting here that the length of time taken for a simulation will be dependent on both the hardware and the software used.

4.4 Commercial software packages used for CFD

Each software package aimed at the CFD market has to assist the user in carrying out the tasks that form the analysis process. This is done by providing, typically, three main pieces of software:

- a pre-processor,
- a solver,
- a post-processor,

together with a variety of utility programs. The use of all these programs will be explained below.

4.4.1 *Pre-processing programs*

All the tasks that take place before the numerical solution process is started are called 'pre-processing'. This includes the first three phases of the analysis process that we have discussed – thinking, mesh generation and flow specification – and the part of the fourth phase that defines the numerical control parameters. Whilst the first phase needs considerable thought, and considerable engineering judgement, if the physical flow problem is to be translated into a problem that is solvable by the CFD software it does not involve any computing. It is only when this first phase has been completed that the computing starts.

To assist in the computational part of the pre-processing phase, most software packages have a pre-processing program that can be used to carry out the following operations:

- Define a grid of points and perhaps volumes or elements.
- Define the boundaries of the geometry.
- Apply the boundary conditions.
- Specify the initial conditions.
- Set the fluid properties.
- Set the numerical control parameters.

In carrying out these tasks the user has to interact with the computer in some way and so the pre-processing program usually has a graphical interface, so that parameters can be set and the resulting changes

75

seen quickly. This is particularly important when the mesh is being built. Also, datafiles can be read that contain lists of commands so that repetitive sets of instructions, say for a similar but not identical problem, do not have to be typed too often.

Usually, the most difficult task in the pre-processing phase is the generation of the grid of points or mesh. Quite often this task can be simplified by using software that has been especially designed to carry out mesh generation. One example of this is the use of programs written to produce meshes suitable for the finite element analysis of structural problems. Such software is commonly available and can interface with computer-aided design systems. This allows the analyst to access computer models of objects, the surface data of which can form the basis for the geometry around which the mesh for a CFD simulation can be built.

4.4.2 *Solving the equations*

Each package has a program that solves the numerical equations for the problem under consideration. This program must be given all the relevant data that has been defined by the pre-processor. To transfer the data between the programs, the pre-processor writes out datafiles that the solver program can read. These files can also be moved, if necessary, between computers. This is extremely useful as it means that the solver program can run on a machine specifically designed for high speed numerical work, such as a supercomputer, while the interactive tasks are carried out on a smaller machine. This splitting of the tasks between machines enables the hardware to be used in the most efficient manner, keeping graphics-intensive and so-called *number-crunching* activities separate.

Once the datafiles are in place, the solver program is activated and the required solution process carried out. At the end of this phase, further datafiles will be available, which may have to be transferred back to the machine where the pre- and post-processing programs are run.

Although the solver program is the core of any CFD software system, the user sees little of its operation.

4.4.3 *Post-processing programs*

As large numbers of points have to be created within the flow domain if reasonable simulations are to be obtained, and as several variables are stored at these points, computer graphics is often the only means of assessing the data written by the solver program. The post-processing program is used to display the results, and, as with the pre-processor, this program is interactive and so is usually run on the same machine as the pre-processor.

Typical pictures obtained with the post-processor might contain a section of the mesh together with vector plots of the velocity field or contour plots of scalar variables such as pressure. These pictures enable global trends in the data to be seen.

4.4.4 *Utilities*

Several utility programs are sometimes provided that do not form part of the above system of software. These programs can be used to convert the datafiles written by one system into a format that can be read by another system. This is common for files containing mesh and results data.

Using these utilities the data can be transferred between engineering software systems and this can be extremely useful if an organization has the use of commercial mesh generation software, such as would be provided with a finite element structural analysis program. These programs can be used to build a mesh that can then be accessed by the CFD analysis system. The files that are transferred are often referred to as *neutral* files, as they can be read using the standard text editors of many systems or by small programs that are written locally.

5 Describing flow problems in engineering

Producing a computer simulation of a flow problem requires the analyst to provide a large amount of data to the solver program. It is the quality of this data, in terms of both suitability and accuracy, that may well determine the quality of the results of the simulation. Because of this, users of CFD software must be very familiar with the flow problems that they wish to simulate. When using computers there is a strong temptation to start computing as soon as possible, but in this case it is much better if considerable thought is given to the problem before starting to use a computer, and so the urge to compute before thinking must be resisted.

As an aside, if you are considering having an analysis undertaken using CFD, then please be aware of the following. At times the analyst will use hard information which will be gleaned from a variety of sources. This sort of information includes much that is not controversial and is well known. At other times, however, the analyst must rely heavily on the experience of running similar fluid flow simulations when deciding how to model the problem. This is because the CFD analysis will sometimes demand information that does not exist, or the software may not model exactly the situation that is required. In such circumstances the quality of the analyst can be crucial to the simulation being successful.

The key to a sound analysis is the production of a specification of the flow problem. This is a clear exposition of the reasons why the simulation is being carried out and of what the physical flow situation is. Once it has been produced it can be translated into the set of data that is required by the simulation package. This chapter looks at how such a specification is built up and then looks at an example of a specification for a realistic flow situation that will be simulated in Chapter 10.

78

5.1 Producing a specification

A specification for a flow problem must be sufficiently detailed that the analyst can obtain from it all the information necessary to define the flow problem to a CFD solver program. This information comes from a good understanding of the flow problem which the analyst must obtain by talking with the people who require the results of the simulation. In particular the analyst must know the following three things:

- why it is that the simulation is required;
- what the geometry of the problem is, in broad terms;
- what the possible flow behaviour might be.

5.1.1 *Knowing what is required of the analysis*

Carrying out the analysis of a fluid flow problem is an expensive business. If someone wants to commission a computational analysis of a flow problem, considerable expense will be involved as access to computer hardware must be achieved, the necessary software must be found and the labour costs in either time or money are not insignificant. Consequently, there must be good reasons for carrying out the analysis and the analyst must therefore explore these reasons first, by talking to the people that need the results of the simulation, such as design engineers. At this stage the analyst should also be able to decide if a CFD simulation will give the required results.

The reasons for an analysis being carried out are many and varied but they often include such things as the determination of the forces and moments on a body so that the motion of the body can be predicted or analysed, the prediction of the pressure throughout a flow or the prediction of the ways in which the fluid moves over or through a system. Sometimes the analyst will have to work out the form of the results that the simulation should produce from a vague description of an engineering problem. For example, the work done in pumping a fluid at a given flow rate through a series of passages of an engineering device might be too great and the reasons for this may not be known. A computational model of this problem would show what happens to the fluid as it passes through the passages and it will also give a prediction of the fluid pressure everywhere in the device. From this information the areas where the fluid pressure is lost can be identified, as this will usually occur where the flow is separated. With this information the computer model could be altered so that a prediction is made of the flow through a modified geometry that should reduce the regions of separated flow. The results of the prediction should show whether the modification of the geometry would lead to a reduction in the pressure losses in the physical flow.

Once the analyst knows the reasoning behind the flow problem it is easier to plan ahead so that the computational model produces the necessary information. One further benefit of this discussion between analysts and their clients is that they get to know each other and their respective problems. Such an understanding can help the analysis process to be brought to a successful conclusion, especially if things do not quite go as planned.

At the end of this initial part of the specification phase the analyst should have a list of the data that the computational model must produce. This could include the change in pressure through a system, the local pressure field, the local flow velocities, the time variation of a variable at a given point or many other pieces of information. Once this list has been compiled an assessment of the suitability of CFD in giving reasonable results should be made. We need to be aware at this stage that CFD cannot produce sensible results for all physical fluid flow problems, and we will discuss why this is in Chapter 10 after we have looked at the results of some simulations. If the analyst concludes that CFD is not a suitable tool to use in obtaining the required results, whatever the reasons for this, then the analyst must highlight these problems to those who want the results and suggest that the analysis is not carried out with CFD. There is no point in running a simulation if it is likely that the results will be of poor quality. This would only frustrate those who need the information and give the use of CFD a bad name. Everyone should always be aware that sometimes it is easier and cheaper to perform a physical experiment rather than a computational one, and sometimes it is more accurate too.

5.1.2 *Specifying the geometry of the problem*

Once the reasoning behind the analysis is known the actual specification of the problem can be prepared. When looking at any flow problem it is important to be able to describe the physical boundaries that contain the fluid. This is particularly important for engineering flow problems where it is usual for at least some part of the boundary to be a man-made object and it is a prediction of the effect of this object on a flow that is required from a CFD analysis.

When we solve the equations governing fluid flow using a computer, we need to have a mesh of points at which the flow variables can be stored, as we saw in Chapter 3. These points have to be created both on and within the bounding surfaces of the flow and so some means of describing the geometry of these surfaces is required.

Various sources of geometrical data are available and these can be used by the analyst to describe the bounding surfaces. For example, this data might come from:

- analytical descriptions of shapes in two dimensions given by such things as points, lines, arcs and splines;
- engineering drawings;
- databases created by computer-aided design (CAD) systems;
- measurements taken from existing hardware.

From such sources most of the bounding surfaces of the flow domain may be determined precisely. When building the mesh of points inside the flow domain we will use this precise information (see Chapter 6), but during the specification stage it is sufficient to know roughly where these surfaces are in relation to each other and how they fit together. A simple sketch might help to show this. It is also worth remembering that when we build the computational model a complete description of the bounding surfaces is required, and that some of these surfaces might not be physical surfaces. For example, the non-physical surfaces could be the flow inlet or outlet or the boundary of an external flow problem that is effectively at infinity (the far-field boundary). These non-physical surfaces will need to be created later, but the sketch should at least draw attention to where they are.

5.1.3 *Defining the flow*

Once the geometry of the problem is understood the analyst must think about the flow itself and try to visualize what is happening to the fluid within the bounding surfaces of the flow. The initial step in defining the flow is to know which fluid is to be studied. This could be air, water or any other fluid, and the values of the density and viscosity of the fluid need to be found. Once the density and viscosity are known a calculation can be made of a parameter known as the Reynolds number. This is a non-dimensional number, often designated by Re, which is defined as

$$Re = \frac{\rho V_{ref} D}{\mu} \tag{5.1}$$

where V_{ref} is a reference velocity such as the inlet velocity and D is a characteristic length which might be something like the length of an object or the width of a duct. This parameter is useful in determining whether a flow will be laminar or turbulent, as we shall see, and is one of a number of non-dimensional parameters that are used in fluid mechanics to characterize flows. We will discuss several more in Chapter 11 when we look at other types of flow.

Next, the production of the main part of the specification can be tackled. As we have already stated the CFD solver must be aware of the boundary and initial conditions that are appropriate for the flow under consideration. The investigation of these conditions can be started by building up a picture of the flow structure that might occur. This is done by thinking about the physical boundaries of the problem that were identified in the previous part of the specification process. From our sketch of the location of the boundaries we should be able to identify those surfaces where the fluid enters or leaves the geometry and those surfaces which are the solid surfaces. This information can then be used to gain some idea as to the flow structure within the geometry. The flow structure might include such things as the direction of the flow, the location of vortices, areas of separated flow, boundary layers and wakes. The existence of these features within the flow can then be added to the sketch that we are building up.

As part of the physical flow structure, areas where the flow variables such as velocity have large gradients, for example in boundary layers and wakes, will be identified and this information can then be used when the mesh is built so that sufficient points are placed within the mesh in these regions. Also, the flow structure will help to identify the type of boundary condition that should be applied to each of the boundaries and the initial state of the flow variables. Remember that it is the flow information on the boundaries of the geometry, the boundary conditions, and the state of the flow variables at the beginning of a time-dependent problem, the initial conditions, that determine the numerical solution to a particular set of equations. By now the sketch should have most of the information about the boundaries on it and this needs to be translated into the form needed by the analysis. This is done by looking at each boundary in turn.

Some common specifications that need to be made at boundaries are as follows:

- to fix the velocity (at an inlet or a wall where the flow is laminar);
- to activate a log-law velocity profile (at a wall where the flow is turbulent);
- to activate appropriate functions for the turbulent kinetic energy and its dissipation rate (at a wall where the flow is turbulent);
- to fix the turbulent kinetic energy and its dissipation rate (at an inlet of a turbulent flow);
- to fix the pressure (at an outlet);
- to do nothing (at a symmetry plane where the velocity gradients normal to the plane are zero);
- to specify a pair of cyclic boundaries where the flow variables are the same at corresponding points on the two boundaries.

If we wish to solve a steady state problem the flow should now, in theory, be completely specified, but if we wish to solve a flow with a time variation which is either real or assumed in the solution procedure, then the initial conditions must also be specified. These are the values of all the flow variables at the start of the calculation and they need to be defined at every point in the flow domain. Often the values are not known exactly and so some sensible values have to be assumed. Even if the problem is to be solved as being steady in time we must sometimes specify some initial conditions. Many programs will assume an initial set of values for the flow variables, but it can help to give a better guess as less computational effort might be used in reaching the final solution.

5.2 An example of a flow specification

So that the above specification process can be illustrated, we will now take a flow situation and consider how a specification can be produced by this process. The example that we will use is that of a two-dimensional slice of the flow of air over a saloon car when it is placed in a wind tunnel. This is one of the examples that will be used as a demonstration example in the chapter of case studies, Chapter 10.

First, we must think about the reasons for carrying out the simulation. Let us imagine that we are working for a vehicle manufacturer and ask ourselves the question, 'What does the company want to find out from the simulation of the flow over a car?' Cars are tested in a wind tunnel for a variety of reasons that include the search for information about the forces and moments on a vehicle that can be used to predict the vehicle's fuel economy, its top speed and acceleration and its response to gusts of wind hitting the vehicle from the side. The data that is extracted from these wind tunnel tests includes the following:

- the drag on the car when the car is at various angles to the flow;
- the lift on the car at the same set of angles;
- the side force on the car at the same angles;
- the rates at which the cooling system of the car can extract heat from the engine;
- the rate of cooling of the brakes.

If we carried out a three-dimensional simulation of the flow around a vehicle we could obtain values for the forces and moments on a basic body shape, but none of the above information can be found from a two-dimensional calculation. This is simply because the two-dimensionality of the calculation will make the results meaningless; however, the procedures are just the same as those used for three-dimensional calculations and so this example can be seen as a reasonable test case to pursue. In Chapter 10 there will be a discussion of the use of CFD in calculating the three-dimensional flow over a vehicle after the two-dimensional calculation has been made.

Let us imagine that we wish to run this simulation to investigate the flow structure around the vehicle, which can give some pointers to the three-dimensional flow. Consequently, we will want to be able to plot the velocity vectors around the vehicle at the end of the simulation. Having decided this we can move to the second step in the specification process, that is, the sources of data for the geometry and the arrangement of the boundaries must be found. For the car the shape might be defined as a set of engineering drawings or as a set of surfaces stored in a CAD system, but the shape of the wind tunnel must also be decided. Most tunnels are comprised of a parallel working section placed between a contraction and a diffuser. To simplify this problem the tunnel can be taken to consist of a straight floor and roof which are placed at the correct elevations relative to the car, a vertical inlet upstream of the car and a vertical outlet some way downstream of the car. This simplification can be made as the main effect of the tunnel on the car is to constrain the flow around it and this is done by the working section immediately around the car. The fact that the working section has been extended away from the car should have little effect on the flow around the car, but it does simplify the computation considerably. In particular, the outlet needs to be far downstream of the car to reduce the influence, on the flow around the car, of the approximate pressure boundary condition that will be specified at the outlet. All this information is summarized in a sketch of the geometry which is shown in Figure 5.1. The shape of the car comes from a set of two-dimensional curves in space that are derived from the three-dimensional data discussed in Chapter 10.

Having specified the geometry the fluid can be defined. In this problem the fluid is air which has the following properties (at a temperature of 288 K and a pressure of 760 mm of mercury):

density 1.225 kg/m^3
viscosity 1.79×10^{-5} kg/m s
kinematic viscosity 1.46×10^{-5} m^2/s

Then the Reynolds number can be found by taking V_{ref} to be the inlet velocity of 28 m/s and the typical length dimension to be the vehicle length of 4.165 m, giving

$$Re = \frac{(1.225)(28.0)(4.165)}{(1.79 \times 10^{-5})} = 7.98 \times 10^6 \qquad (5.2)$$

From this calculation we can assume that the flow will be turbulent as the Reynolds number is so high.

Now the boundaries of the problem can be analysed and from Figure 5.1 it can be seen that the boundaries can be listed as follows:

- the car surface,
- the tunnel floor,
- the tunnel roof,
- the tunnel inlet,
- the tunnel outlet,

and each must be considered in turn.

The effect of the car surface is simple to understand. At this boundary the flow will be turbulent and the surface will retard the flow. Boundary layers will be created on the vehicle surface. In terms of boundary conditions a log-law profile condition for the velocity will have to be imposed here together with suitable conditions for the

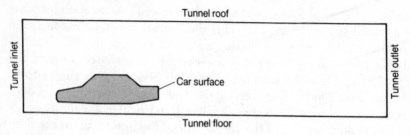

Figure 5.1. Car in a wind tunnel

variables of the turbulence model. Similarly, the tunnel floor will act in the same way and will require similar boundary conditions to be imposed. On all these surfaces the mesh will have to be built such that several points are placed near to the surface in a direction normal to the surface.

The tunnel roof is an interesting boundary in that it will act like the tunnel floor and have a boundary layer on it. However, as it is some way from the car, this boundary layer is unlikely to have a major effect on the flow over the car and so the roof can be taken to be a symmetry boundary so that no flow goes through the surface. This is of benefit to the simulation as the mesh does not need to be very fine near a symmetry boundary, whereas it does need to be fine where there is a boundary layer so that the variation in velocity near the solid surface is captured. By making this approximation for the roof the number of mesh points in the domain can be reduced.

At the tunnel inlet the fluid enters the domain in the horizontal direction at a speed of 28 m/s and so, as both the magnitude and direction are known, the velocity can be specified there. Being carried in with the flow is a natural level of turbulence and this must be specified at the inlet as well. However, at the tunnel outlet, we do not know the speed of the flow at all positions as there is a boundary layer on the floor of the tunnel and a wake behind the car that is generated by the boundary layers on the vehicle's surface. We can deal with this boundary by assuming that the velocity does not vary in the horizontal direction at the outlet and so the derivative of the velocity in the horizontal direction is zero. Further, we can impose a fixed pressure at the outlet as it is sufficiently far from the vehicle that this boundary condition will not affect the results we want to obtain. Normally we set the outlet pressure to zero.

In terms of the flow structure the analysis of the boundaries gives

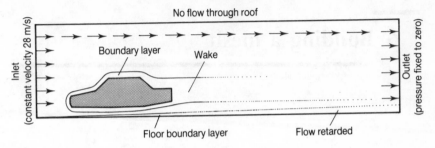

Figure 5.2. Flow features in the tunnel

us a picture of what is happening in the flow. Fluid enters through the inlet and is retarded by the tunnel floor and the vehicle surface, forming boundary layers there. From our knowledge of fluid flow we know that the flow will separate somewhere towards the rear of the vehicle forming an area of fluid that has a reduced speed behind the vehicle. This is the wake of the vehicle. At the roof of the tunnel the flow is constrained to move horizontally, and the fluid leaves the domain at the outlet. All this information can be added to the sketch, as shown in Figure 5.2.

As a last step we must decide upon the initial conditions. For this problem a sensible way to approach things would be to set the horizontal component of velocity to the speed of the inlet velocity, to set the vertical component of velocity to zero and the pressure to zero. Turbulence values can be set to be the inlet values as well. Now the specification is complete and we can turn to the building of the actual computational model.

6 Building a mesh

Once the specification of the flow problem is known we can turn our attention to building a computer model. The first part of this is to build a mesh of points throughout the flow domain and perhaps produce the necessary volumes or elements. When modelling a simple problem this process takes very little time, but when modelling a complex problem such as the flow inside a series of passages, say the coolant flow in an internal combustion engine, the process can take several man-months to complete. Often it is this phase of the analysis process that determines the total time required to obtain results from a simulation, as all the other phases, including the actual computation of the results, can be carried out quite quickly. Similarly, *the overall cost of the analysis* can be *totally dominated* by the costs of the labour required to build the mesh.

In this chapter we will discuss the reasons for building a mesh, the requirements that a mesh must satisfy if it is to give satisfactory solutions and the types of mesh that can be built. Then we will discuss how a mesh can be built by using a variety of software tools. Finally, we will look at ways in which a mesh can be modified in the light of the results of a flow simulation such that better results are achieved.

6.1 The need for a mesh

In Chapter 3 we looked at various ways of discretizing the governing partial differential equations of fluid flow so that numerical equations were produced. Regardless of which of the three discretization techniques is used (the finite difference method, the finite element method or the finite volume method) a mesh of points has to be produced within the volume of the fluid. This can be considered as the discretization of the space in which the flow takes place. If we use the finite difference method then the values of the variables at the points are used to produce equations for the variables that enable a solution to be determined. This involves a grid of points. However, if we use the finite volume method then the points are arranged so that they can be grouped into a set of volumes and the partial differential equations can be solved by equating various flux terms through the faces of the volumes. Also, if we use the finite element method then the points are grouped to define elements within which the numerical analogue to the partial differential equations can be set up. In both the latter cases the structure of the mesh does not depend on the discretization method.

As a consequence of this we can see that although we need a mesh to solve CFD problems regardless of which of the three discretization techniques has been used, the mesh itself will be influenced by the discretization technique. This is not the only influence as the expected variation of the flow can also have an effect on the way in which the mesh is built.

6.2 Creating a mesh for a given flow

Every flow problem will contain a wide variety of flow features in the domain. That is, things such as vortices, boundary layers, regions of rapid fluid velocity and pressure change and separation regions occur, and all of these need to be modelled by the CFD simulation. If we are to have a mesh that is capable of modelling these features, where the gradients in space of the flow variables are high, then we must be aware of where these features might occur. This shows the importance of the sketches that we developed as part of the specification process, as these can be used to highlight the positions of the critical regions in the flow.

In the critical regions we need to have a large number of points within the mesh. To see the reason for this we must refer back to Chapter 3. There we saw that all the numerical methods assume that the flow variables vary in some simple way between the points or within an element or volume. This variation is usually linear but, for finite element codes, a quadratic or even higher-order variation is sometimes used. Consequently, if the flow varies rapidly in space, as it does in the critical regions of the flow, a fine grid will be needed to describe the variation accurately.

We can see this clearly in Figure 6.1 where a one-dimensional variation in a variable U is assumed to occur in the x-direction. Let us

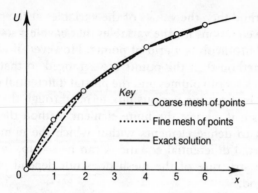

Figure 6.1. Effects of mesh density

assume that some numerical method has given us a set of values for U which is exact at a number of points in the x-direction. This will never happen in practice but it is the best that a numerical method can do. If we take the numerical prediction of U to be the straight lines between these points, then several sources of error in the variation can be seen. First, if the values are obtained at only a few points, which we will call a coarse mesh, then the solution is not an accurate representation of the variation. We can see that in the region of $x = 1$, the numerical value of U is too small and, in the region of $x = 0$, the numerical approximation to the derivative dU/dx is too small. If we know the values of U at more points, that is, on a finer mesh, then we can see that numerical description of the variation is much more accurate.

This is extremely important as we must have accurate values of the variables and their derivatives if we are to simulate the governing equations accurately. Any error in either the variable U or its derivative dU/dx can lead to the numerical solution of the equations being in error. A typical example of this is that flow separation on the surface of an object may not be predicted if the mesh is too coarse near the surface of the object. An example of this is given in Chapter 10.

Even though we know that the mesh must be very fine in the critical regions we still have the problem of knowing where these regions are and how fine the mesh should be. Along solid surfaces there will be a boundary layer and so there must be several points close to the surface in a direction normal to the surface. This allows the numerical solution to model the rapid variation in velocity through the boundary layer. This is an example of the geometry of the domain influencing the way in which the mesh must be built. Another example is where a surface has a large amount of curvature causing a rapid variation in pressure in the flow direction. However, large flow gradients also exist in areas of the flow away from the solid surfaces, say in the wake of an object or, if we are modelling a compressible flow, near a shock wave. Creating a suitable mesh in these areas is more difficult as the exact location of the critical areas is difficult to determine. One way of proceeding is to assume the position of the critical areas and build a mesh taking this into account. Then, once the simulation has been run, the actual results of the simulation may help us to determine the actual positions of the regions of high flow gradients. So we see that information obtained from the results of a simulation can be used to modify the mesh and the technique is known as *adaptive meshing*. We shall discuss this further in Section 6.5.

6.3 Mesh structures

6.3.1 *The basic parts of a mesh*

Given that a mesh must be suitable for the discretization technique and also for the flow, we will now look at the different types of mesh that can be built. A first step in this is to determine what the basic parts of a mesh are. From our discussions in Chapter 3, we already know that a finite difference mesh will consist of a set of points, that a finite volume grid will consist of points that form a set of volumes and that a finite element mesh will consist of sub-domains known as elements on which the variables are found at fixed points known as nodes. The following, then, are the basic parts from which meshes are built:

- points, sometimes called nodes,
- volumes, also known as cells in some documentation,
- elements,

but which of these parts are needed for a mesh depends on the discretization method being used. In all the discussion that follows we will use the terms 'volume', 'cell' and 'element' to mean a sub-domain without implying that a particular discretization technique is being used.

Various mesh structures which are made up of these parts can be built and we shall look at this in the next section, but it is useful to

91

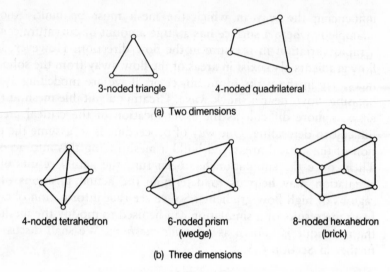

Figure 6.2. Some common sub-domains

note here the range of sub-domains, be they volumes or elements, that are used. In structural finite element programs a wide range of element types can be used and these are classified by the shape of the sub-domain and the placement of nodes in the domain. With CFD programs a much more restricted set of volumes or elements is available at present. By far the most common volume or element, for use in three-dimensional meshes, is a hexahedron with eight nodes, one at each corner, and this is known as a 'brick element' or 'volume'. For two-dimensional applications the equivalent element is a four-noded quadrilateral. Some finite volume programs have now been released which have the ability to use tetrahedra in three dimensions or triangles in two dimensions. Most finite element CFD codes will allow these elements to be used together with a small range of other element types. Figure 6.2 shows some of the commonly used sub-domains.

6.3.2 *Types of structure*

Now that we know what the constituent parts of a mesh are, we can think about how to arrange them through the domain. This arrangement is known as the *structure* of the mesh or the *topology* of the mesh. When using the finite difference method the points are the

92

positions in space where the variables are calculated and they are arranged in what looks like a grid of cells. In contrast to this, when using the finite element method, the points are the nodes of the set of elements used to split up the fluid volume, and the elements can be arranged in any way, provided that the faces of the elements are aligned correctly. We are able to do this as the calculation on any one element requires information from that element alone. The interaction between the elements takes place when the element equations are added together to form the global equations. With the finite volume method the actual implementation of the numerical solution will determine which scheme of volume placement can be used. Some programs demand that the volumes are placed in the same way as they would be for a grid of finite difference cells and others allow a finite element-like placement.

From this we can see that there are two ways in which the mesh structure can be arranged. These arrangements are as follows:

- *A regular structure or topology*, where the points of the mesh can be imagined as a grid of points placed in a regular way throughout a cuboid (also known as a 'shoebox'). These points can then be stretched to fit a given geometry and this is shown in Figure 6.3. Note that when the mesh is stretched the connections between the points do not change. The stretching takes place as if the mesh were made of rubber, and the

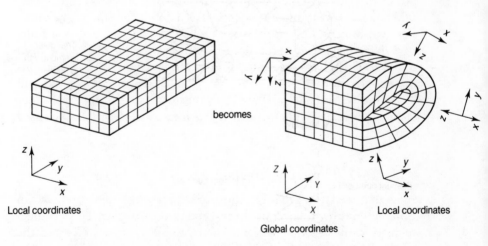

Figure 6.3. Transformation of a mesh with a regular structure

so-called topology, or form, of the mesh remains the same. Consequently, if we consider any point in the mesh it will be connected to the same neighbouring points both before and after the stretching process. Sometimes these meshes are called *structured meshes* as they have a well-defined structure, or *mapped meshes* as they can be seen as a cuboid mesh that has been mapped onto some other geometry. When considering these meshes it is useful to think of a *local coordinate system* within the mesh. This enables the orientation of the cells relative to each other to be determined, and so before the mesh is transformed the axes of this system are the edges of the cuboid. Once the transformation into the actual coordinate system being used, the *global coordinate system*, is carried out the local coordinate system axes become dependent on the position within the mesh. This is shown in Figure 6.3.

- *An irregular structure or topology*, where the points fill the space to be considered but are not connected with a regular topology. Figure 6.4 shows a two-dimensional example of this type of mesh formed with triangular elements. Note that the cell faces do not overlap. We can see from the magnified section of the mesh that element number 1 has the three nodes labelled a, c and d at its corners, and that element

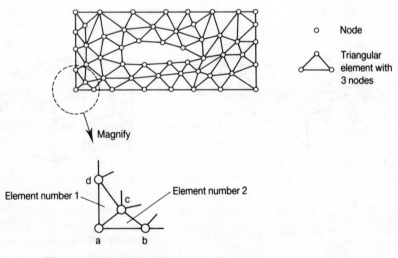

Figure 6.4. A mesh with an irregular structure

94

number 2 has the nodes labelled a, b and c at its corners. The fact that any particular node is attached to an element cannot be known from the form of the mesh, and so a numerical table must exist that describes the arrangement of the mesh by listing which nodes are attached to each element. This contrasts with the regularly structured mesh where a knowledge of the location of a cell within the mesh enables the labels of the points at its corners to be found implicitly. A mesh with an irregular structure is often referred to as an *unstructured mesh* or a *free mesh.*

Relating the mesh structure to the numerical method, finite difference programs require a mesh to have a regular structure and finite element programs can use a mesh with an irregular structure. In theory, finite volume programs could use a mesh with an irregular structure, but many implementations insist that the mesh has a regular structure.

As we mentioned in Chapter 3, when a mesh with a regular structure is used there is an advantage in that the solver program should run faster than if a mesh with an irregular structure were used. This is due to the implicit relationship that exists between the number of a cell or a point and the numbers of its neighbours in a regular mesh, which enables data to be found easily. No such relationship occurs for meshes that have an irregular structure and so when trying to find the values of flow variables in neighbouring volumes there must be a computational overhead. This often takes the form of a look-up table which relates the faces to the cells or the nodes to the elements.

Many flows that are of interest to engineers take place in or around the complex geometries whose boundaries are man-made objects. With some ingenuity on the part of the analyst, it is possible to fit a mesh with a regular structure to some of these geometries, but with many geometries this is not possible. This is where meshes with an irregular structure can be used to great advantage, as these meshes can be used to describe the most complex of geometries due to there being no restriction on the structure of the mesh. This can make the mesh generation process much easier and in some cases it is a prerequisite for producing a simulation. Another advantage of using irregularly structured meshes is that they can be created by automatic mesh generation algorithms, some of which are described in Section 6.4.5. These algorithms generate meshes which are unstructured, using elements such as tetrahedra.

With some CFD programs it is possible to have several meshes which have a regular structure combined together. Programs use these meshes in an attempt to gain the speed advantage that comes from using a regular mesh whilst retaining the flexibility to model complex geometry. This combination of meshes is called a *multi-block* mesh as it can be seen as a series of blocks built together. There is, of course, a restriction on the way that these meshes are built to ensure that the cell faces do not overlap at block boundaries.

As a final point on the structure of a mesh it is worth mentioning two terms that are often met when dealing with regular meshes. In Chapter 3 we looked at some examples where each of the discretization techniques was used. In these examples the domain geometry was simple and the partial differential equations were discretized directly in terms of the Cartesian coordinates. When meshes are built for more complex geometries, the partial differential equations are sometimes transformed into a general cell-based coordinate system. This is especially true when dealing with finite difference methods and finite volume methods which require a regular mesh, as the local coordinate system can be used. This transformation of the equations enables a regular mesh to be used even though it is not rectangular. In some transformations of the equations the mesh of points is required to be *orthogonal*, which means that the sub-region faces must meet at right angles to each other. If these meshes are used fewer terms are required to produce the transformation of the partial differential equations and so less computational effort is required to compute the solution. If the mesh is *non-orthogonal*, then the extra terms have to be programmed and the solution requires more computational effort. Sometimes programs which should use an orthogonal mesh can be run with non-orthogonal meshes but the results that are produced are less accurate.

6.4 Building meshes

6.4.1 *Defining the geometry*

In the specification stage of the process that we discussed in Chapter 5 we saw that we need to determine the sources of geometrical data and to produce sketches of the positions of the bounding surfaces of the flow domain. Now we must use the sources of geometrical information, be they sketches or engineering drawings or computer models, and ensure that we can find the location of the bounding surfaces in terms of the coordinates, say x and y and, possibly, z.

For two-dimensional problems we can create the bounding surfaces using points to define a series of lines and curves. These curves might be defined as circular arcs, simple polynomials or splines. All of these constructions are described by equations that define the relationship between the coordinates of points that make up the curve. For example, we all know that a line can be described by the relationship

$$y = mx + c \qquad (6.1)$$

where m is the gradient of the line and c is the value of y when x is zero. By substituting for the gradient in terms of two known points on the line, equation (6.1) becomes

$$y = \frac{y_2 - y_1}{x_2 - x_1} x + c \qquad (6.2)$$

where the subscripts refer to the two known points. These equations describe a line which is infinite in length, but we will only use lines of finite length to describe the geometry of the flow domain. This means that we need to know the end-points of the line.

Similarly, a circle can be described by

$$(x - a)^2 + (y - b)^2 = r^2 \qquad (6.3)$$

where the centre of the circle is at $x = a$, $y = b$ and the radius is r. A part of this circle is a circular arc and three points can be used to define it. Usually these points are taken to be the two end-points of the arc and a point on the arc somewhere between them. This enables both the limits of the arc to be defined and the unknown constants in equation (6.3), namely a, b and r, to be calculated.

Splines are more complex curves, but they are also defined by points in space. Usually four or more points are used, but they do not have to be on the curve itself. Note that a hierarchy is being formed here in terms of the numbers of points required to define a curve. Two points define a line, three points define an arc and four points or more define a spline.

In three-dimensional problems the geometry might be defined by similar three-dimensional constructions in the form of a so-called *wire-frame model*. In these models the edges of each surface are defined and, sometimes, the form of the various surfaces is known as well. Often these surfaces have a simple form such as a plane or part of a sphere or cylinder, but they can also have a more complex form. Another type of computer model is known as a *solid model*, where the computer stores not only geometric information but things such as the mass of an object. When using a solid model the geometry of an object is defined in ways similar to those used by wire-frame models.

Typical ways of describing the more complex surfaces are as follows:

- *Numerous simple patches*. Here the surface is discretized into a series of patches which are usually triangular or quadrilateral in form. This is the way that a surface would be described by the faces of a mesh of linear elements or cells.
- *Coons patches*. These are patches over which the coordinates of points on the surface are determined from the bounding curves alone. Consequently, once the boundaries of a surface are determined the surface itself is defined. Three or four curves in space which form a closed loop are often used to define the boundaries. Note that an infinite number of surfaces will be able to fit through a given set of boundaries but the Coons patch description defines only one surface. The assumption is made that the patches are sufficiently small so that a good approximation to the surface is given. This can lead to problems if a surface is highly curved and only a few Coons patches are used to model it. In this situation each patch will be too large and the surface definition will not have the required curvature.
- *Bezier surfaces*. These are surfaces which are described by a set of Bezier polynomial curves. Each curve is defined by four points, the two end-points of the curve plus two interior points which need not be on the curve. By moving the two interior points the curve can be manipulated to have a wide range of shapes. Bezier surfaces give an improved description of a surface when compared to a Coons patch description as information from within the boundaries is used to define the surface. This helps to lock a surface in space and so the number of surfaces that could fit the description is reduced. These surfaces were developed for Renault, the French vehicle manufacturer, as they had a requirement for computational surfaces that could be manipulated interactively when modelling new vehicles in the styling studio.
- *Non-uniform rational B-spline surfaces (NURBS)*. These are

98

similar to Bezier surfaces, but the curves that are used to define them are based on different points to the Bezier curves. The end-points of the curves are only approximated, but the points that are used to define the polynomials ensure that the first- and second-order spatial derivatives are continuous at the end-points.

Such is the complexity of these curves and surfaces that a computer has to be used to manipulate the data. For our purposes of building a mesh for use in a CFD analysis, it is not necessary to understand the mathematics behind the descriptions, but the analyst should have some knowledge of the variety of types of surface that exist. For those who are interested, several books describe the ways in which these computer descriptions of objects are handled in computer-aided design (CAD) systems (Encarnacao and Schlechtendahl, 1983; Hordeski, 1986; Bezier, 1986; Rooney and Steadman, 1987).

When we know that the geometry data exists we can start to build the mesh as we know that we can find the coordinates of any point on the bounding surfaces of the domain.

6.4.2 *Determining the mesh structure*

Having made sure that the geometry description is complete the next step is to decide on the type of mesh structure that will be used. Sometimes this might be dictated by the CFD software that is to be used, as some programs only allow a certain structure, or the structure might be decided by the geometry of the domain. As a mesh with a regular structure is simpler to create and should enable the CFD solver to be computationally more efficient, we might attempt to fit, mentally at least, such a mesh to the geometry. If this fails then we must use a mesh with an irregular structure. Although this will lead to some extra work, the effort can be reduced by trying to build a mesh that has a regular structure for much of the domain, only using an irregular structure where absolutely necessary.

Having decided on the mesh structure, a mesh layout can be determined and an estimate made of the number of cells that will be required. To do this requires considerable user-experience, and both the layout and number of cells will depend on the flow that is assumed to take place within the domain.

Once these preliminaries are finished we can think about actually building the mesh. First we must decide upon the means of creating the mesh, and this will depend upon the software tools that are available to us. For simple geometries a short computer program could be written to produce a mesh, but often we will need to find a solution to flow problems in more complex geometries. Some CFD packages have a mesh generator built into the pre-processor program and this may well be suitable for some problems. Also, there are other commercial packages that can be used. Usually, these will be commercial finite element pre-processors, but other programs do exist as we shall see later. It is important to note that every organization will have different tools available, and the analyst must find out what these are.

6.4.3 *Building a simple mesh with a regular structure*

Many problems can be solved by using a mesh that has a simple regular structure. This is made easier by the fact that many CFD packages, if they require a mesh to have a regular structure, allow some cells to be declared as what are known as *dead cells*. This is an extremely useful feature that enables a variety of blocks of regular cells to be used to model some complex geometries. For example, the car problem described in Chapter 5 can be thought of as nine two-dimensional blocks arranged as shown in Figure 6.5. The flow domain consists of blocks 1, 2, 3, 4, 6, 7, 8 and 9, and block 5 is obviously inside the vehicle surface. It is, therefore, the cells in block 5 that are declared as dead cells. It should be noted that even when dead cells are declared the appropriate number of cells must be created as it is the existence of these cells that keeps the book-keeping of the analysis program correct. This book-keeping is essentially the management of the data storage in arrays, and leads to efficient solutions.

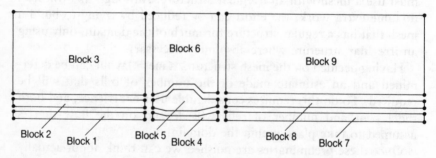

Figure 6.5. Mesh blocks for a two-dimensional car

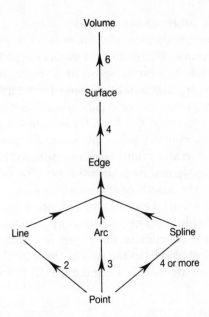

Figure 6.6. A hierarchy of entities

One common way of producing a regular mesh on each block is to use the hierarchy of entities as shown in Figure 6.6. In this diagram we consider the hierarchy for four-noded two-dimensional cells or eight-noded three-dimensional cells. We can see that at the bottom of the hierarchy is the basic geometrical entity which is a point, several of which can be linked to form lines (from two points), arcs (from three points) or splines (from four points or more). By combining adjoining lines, arcs and splines the third level entity, the edge, can be created. If four edges form a closed loop they can be seen to be the boundaries of a surface, and six surfaces can be used to bound a volume. This set of relationships is determined by the elements being considered as, once the surfaces for a two-dimensional problem or the volumes for a three-dimensional problem are defined, the cells can be formed. This is done by mapping the surfaces into a square, and by mapping the volumes into a cube. These squares and cubes are used to define a local coordinate system in which the cells can be created before being transformed back to the global coordinate system which defines the real domain. Whilst commercial software packages use such a hierarchy to produce a mapped mesh, often it is useful to think in terms of this hierarchy even if the mesh

101

is to be produced by some other means such as a simple computer program.

Returning to the car example, Figure 6.5, each block can be seen to be a surface with four edges. The mapping of a mesh onto this surface is fairly straightforward and will be discussed in Chapter 10. The mesh can be created with the cells unevenly spaced so that more cells are placed in critical regions of the flow. This is done by using a geometrical progression to bias the positions where the points are created. Such a progression creates points with the distance between neighbouring points being governed by a simple ratio. For example, each cell may be 0.8 times the length of the previous one.

Finally, before creating a mesh, the node and cell numbers have to be calculated within each block. This is done by allocating the number of cells in the local coordinate directions for each block, remembering that the number of cells in the two or three local coordinate directions must be consistent with a regular mesh. For the car

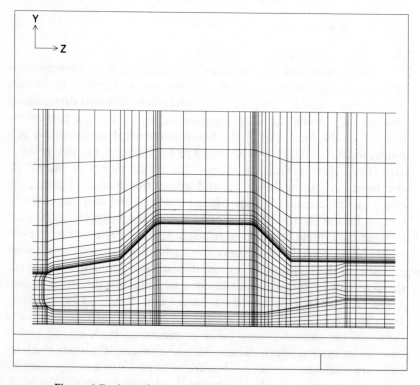

Figure 6.7. A mesh around a car

example, a sensible number of cells might be as shown in Figure 6.7 where the cells within each block are shown for a region close to the vehicle surface. This mesh has been created by placing several cells in the direction normal to the solid surfaces so that the boundary layer region can be predicted more accurately.

6.4.4 *Using commercial mesh generation software*

Commercial mesh generation packages have been around for some time now and are aimed at the finite element structural analysis market. The meshes that they produce can also be used for CFD calculations, provided that they are built with the special requirements of the CFD solver program in mind. As structural finite element work involves many different element types and the use of many materials, these mesh generation packages are extremely general in their capability. This generality is a great strength but it can also make the packages slow to use for CFD applications, as the database of the package is often very large and so is stored on the secondary storage media of a system, usually disks. This can make data access slow. Another inefficiency can arise when the programs ask for information that is not relevant for CFD applications. Often this involves the definition of material properties for every block in the mesh. In structural calculations the properties may well vary from block to block as different materials are used, but in flow problems it is usual to have only one fluid.

Commercial mesh generation packages usually have the following components:

- *A geometry creation routine*, where two- or three-dimensional geometrical data can be created in the form of points, lines, arcs, splines and, sometimes, surfaces. An interface to extract similar data from CAD systems is a common feature as well.
- *A domain definition routine*. This allows the creation of surfaces, in two dimensions, or volumes, in three dimensions.
- *A mapped-mesh generation routine*. This enables a mesh with a regular structure to be created within the domains. These domains must be topologically consistent with the element type being used. For example, if four-noded quadrilaterals are being used to mesh a two-dimensional domain, then a four-edged domain must be used.

- *A free-mesh generation routine.* This enables a mesh without a regular structure to be created within the domains. In this case there is no restriction on the form of the domains, and so they, can be either surfaces bounded by any number of edges (for two-dimensional problems) or volumes enclosed by a set of these surfaces (for three-dimensional problems).

When using commercial mesh generation software, hierarchies such as that shown in Figure 6.6 are used. Usually, this does not cause a problem, but there is one area where errors in the modelling of a geometry can occur. Coons patches are an obvious choice for defining the geometry of a surface within the hierarchy, as the edges are used to define the surface. As we discussed in Section 6.4.1 such a representation of a surface may not be adequate if the patch is too large for the curvature of the surface. One way of overcoming this problem is to define smaller surfaces, but this involves much more work on the part of the analyst. Another way is to use more accurate surface descriptions, say Bezier surfaces or NURBS, derived from a CAD model of an object. Many commercial finite element pre-processors can read these more accurate surfaces from the database of a CAD system. Then, a set of edges can be used to define a Coons patch surface. Once this has been done, the user can tell the pre-processor to calculate the mesh points on this surface by first calculating the coordinates of the points on the Coons patch and then recalculating the coordinates so that they are positioned on the more accurate surface.

As has already been stated, when using mesh generators aimed at producing meshes for finite element structural analysis problems extra information has to be provided during the mesh generation to define the structural properties of the elements as they are created. This not only slows down the mesh generation, but it also means that we have to be selective when extracting the data required by a CFD analysis. At this stage in the CFD analysis process all that is required is a simple definition of the mesh that can be read by the CFD pre-processor. This minimum set of information is restricted to the following two items:

- a list of the positions in space of all the nodes in the mesh; usually this will be a list of x-, y- and z-coordinates;
- a list of the element numbers, together with their type and the numbers of the nodes that are attached to them. This is known as the *connectivity list*.

Most mesh generation packages can write this data to a file which has then to be read by the CFD pre-processor. Some pre-processors will read the mesh information file from a small number of the most common commercial mesh generators. If the pre-processor does not do this, then the data has to be translated into a suitable format. This is done by a small computer program which must be written by, or for, the analyst. It is worth noting that each pre-processor reads the mesh data in a different format, and that this can depend on the needs of the software. For example, programs that only use a regular mesh need only read the nodal coordinates, provided that they are given in a pre-defined order, as the connectivity list is implicit in the regular mesh structure. Conversely, programs that can use an unstructured mesh will read both the nodal data and the connectivity list in their own pre-defined format.

6.4.5 *Some automatic mesh generation algorithms*

For simple geometries it is easy to see how a mesh can be built, but when the geometry becomes more complicated the meshing process is more difficult. Several techniques have been developed that can take complex two- and three-dimensional geometries and then automatically produce a mesh that models the geometry. Typically, the mesh will have an irregular structure. As we said at the beginning of this chapter, mesh generation is a costly part of the CFD analysis process because of the large amount of manpower that can be required to build the mesh for a complex geometry. Any savings in the time taken to build a mesh could make CFD a more attractive solution for some engineering design problems, and so these automatic mesh generation techniques are being actively researched.

The first method that we will discuss is *Delaunay triangulation* (Cavendish *et al.*, 1985; Holmes and Lanson, 1986; Watson, 1981). Figure 6.8 shows this algorithm at work for a two-dimensional case where triangular elements are to be created. The algorithm is easily extended to three dimensions where tetrahedral elements would be formed. In Figure 6.8(a) we can see that the basic technique is started by producing nodes on the boundary of the domain and nodes inside the boundary. In this case there are twelve nodes on a square boundary and one node inside the domain. To ensure that the final triangulated mesh has no gaps in it, three extra nodes are then created that define a super-triangle. From Figure 6.8(b) we can see

105

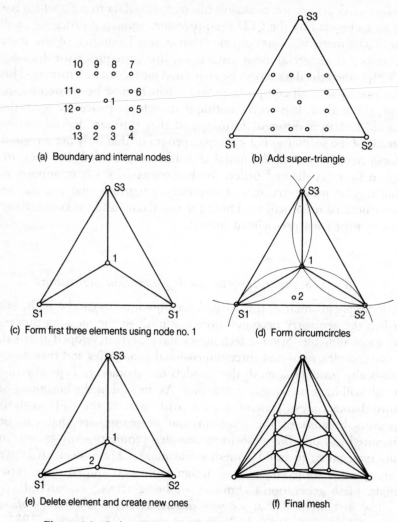

(a) Boundary and internal nodes

(b) Add super-triangle

(c) Form first three elements using node no. 1

(d) Form circumcircles

(e) Delete element and create new ones

(f) Final mesh

Figure 6.8. Delaunay triangulation

that these extra nodes have to be placed so that they define a super-triangle which encloses all of the original nodes of the problem. This super-triangle is taken to be the first element and then one of the original nodes is used to split this element into three new elements (Figure 6.8(c)). Now an iterative element creation procedure can begin. One by one each of the remaining nodes is considered and the mesh modified. To do this a circle is created for each element such that it passes through each of the three nodes of that element. Looking at Figure 6.8(d) we can see the circles of the elements and we will consider node number 2. This node lies outside two of the circles and inside the other. The triangulation algorithm states that if a node lies inside a circle then the element that the circle is attached to should be deleted. Once all the necessary elements have been deleted, new elements can be created that include the node being considered. This is shown in Figure 6.8(e) where the lower element of Figure 6.8(c) has been deleted and three new elements have been created which are joined at node number 2. Then another node is considered and the process continues. Eventually, a final mesh is created such as that in Figure 6.8(f). This can then be modified so that only the original domain, in this case the square, is modelled. This is done by deleting all the elements which are attached to the nodes that formed the super-triangle. Finally, the shape of the remaining elements is checked and, where necessary, the elements are modified to be as near to equilateral triangles as possible. This produces a mesh which does not have elements with a very distorted shape as these elements could cause numerical problems when the solver is run.

The second method is based on the use of the *Quadtree* and *Octree* methods (Yerry and Shephard, 1984; Cheng *et al.*, 1988). These methods take a domain and place it inside four squares if it is a two-dimensional problem, or eight cubes for a three-dimensional problem. These are then sub-divided until the required definition is acquired. Hence the name Quadtree refers to the structure of the elements in two dimensions and Octree refers to the three-dimensional method. Looking at Figure 6.9(a) we can see an example of a two-dimensional domain that is to be meshed. Four squares are placed over the domain, as shown in Figure 6.9(b), and a node created where the squares are joined inside the domain. Each square can then be sub-divided into four more squares and more internal nodes created. Two further sub-divisions are shown in Figure 6.9(c) and (d). Once the element size for the bulk of the mesh is small enough, only the elements that cover the domain boundary are sub-divided. This selective sub-division is shown in Figure 6.9(e), where the shaded circles denote the nodes that are external to the domain but are attached to elements that cover the domain boundary, and the other circles denote internal nodes. This selective sub-division can be continued as required, but it leaves a mesh that is a stepped representation of the domain. To overcome this the external nodes

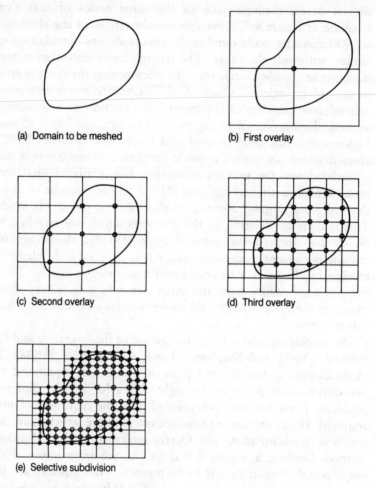

(a) Domain to be meshed

(b) First overlay

(c) Second overlay

(d) Third overlay

(e) Selective subdivision

Figure 6.9. Quadtree method

are moved so that they are on the surface of the domain. Finally, triangular elements can be created to link all the nodes.

Specialized software is available to perform mesh generation using forms of Delaunay triangulation and Quadtree/Octree methods, but commercial finite element mesh generation software can also be used to generate a mesh with an irregular structure in an automatic way. This is often done by meshing the surface of the domain, using triangular elements. Then the volumes that have been defined by the surfaces can be meshed using tetrahedral elements formed from the

108

elements on the surface. At first sight this might appear to restrict such free-mesh generation methods to only using tetrahedra. These can, however, be easily converted to eight-noded brick elements, as shown in Figure 6.10. There, a single tetrahedral element is taken and new nodes formed at each of the mid-sides of the element edges, at the centroids of each face of the element and at the centroid of the whole element. These can then be joined as shown to produce four eight-noded brick elements.

All of these techniques are still in their infancy when it comes to their application to CFD problems. Much research has still to be done before meshes that are suitable for CFD problems can be built

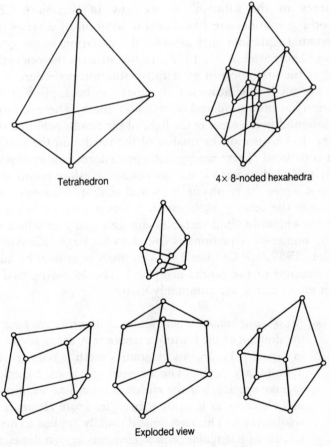

Tetrahedron   4× 8-noded hexahedra

Exploded view

Figure 6.10. Transforming a tetrahedral element

quickly and with a minimum of user intervention. Some people would even claim that this situation will never be achieved.

6.5 Modifying an existing mesh to give a better solution

Once a mesh has been built it is possible to modify it in such a way that the CFD solution that is produced on the modified mesh should be a better one. This modification can take place either before a solution to the flow problem is found or afterwards. Some CFD pre-processors can take a mesh with a regular structure and smooth it, such that the cells form an orthogonal mesh. This can reduce the computing effort required to produce a solution and increase the accuracy of the solution, as we saw in Section 6.3.2. These smoothing routines are based on the solution of a series of partial differential equations that describe the variation in the grid coordinates (Thompson *et al.*, 1982). In this process the original mesh is used as the first guess in an iterative solution procedure.

Other mesh modification techniques can be applied after a CFD solution has been produced on an initial mesh. These techniques are used to modify the mesh in the light of the results achieved on it, and so the dependence of the quality of the results on the user's experience is reduced. These modification procedures require that an initial analysis is made using a crude but realistic mesh of points in the flow domain. From the results of this initial analysis the mesh is recreated such that the density of the mesh points is greatest in areas of the domain where the fluid variables change rapidly or where the error in the numerical equations is found to be large (Zienkiewicz and Taylor, 1989, see Chapter 14). The mesh is said to be adapted to take account of the results generated. The following two types of mesh modification are commonly used:

- *Mesh enrichment*, where additional points are placed within the domain at the locations where they are needed, as shown in Figure 6.11. In this diagram a mesh is required to model a boundary layer. The original mesh of triangles has a regular spacing but the enriched mesh has additional nodes and elements in it so that there are more elements near the solid surface. This technique is usually applied to meshes that consist of triangular cells or elements in two dimensions and tetrahedral cells in three dimensions. Such meshes allow

110

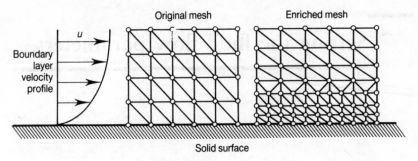

Figure 6.11. Mesh enrichment

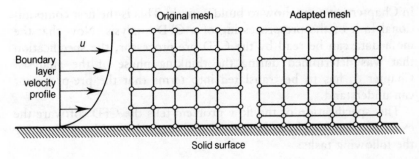

Figure 6.12. Mesh adaption

additional points to be created in the mesh and then the Delaunay triangulation method, or a similar method, can be used to create a new set of elements.

- *Mesh adaption*, where the topology of the mesh stays the same but the mesh points are moved so that the density of points increases where required as shown in Figure 6.12. Here, a boundary layer is again modelled. Note that the number of nodes and elements remains the same in the adapted mesh. Only the node positions are changed. This movement of the points can be brought about by using modified forms of the partial differential equations that are used in some grid generation methods, as was discussed at the beginning of this section.

By using these smoothing or adaption techniques the accuracy of the solution can be increased, but there is a penalty in that extra computational effort is required.

7 Setting the fluid flow parameters

In Chapter 6 we saw how to build a mesh. This is the first computational task of the pre-processing of a CFD analysis. Now that the mesh data can be read by the CFD pre-processor, the specification that was determined during the thinking phase of the analysis, Chapter 5, has to be translated into terms that the pre-processor can understand.

This specification of the flow problem tells the CFD software the exact problem that is to be solved, and it is achieved by performing the following tasks:

- specifying the fluid properties such as density and viscosity;
- determining which flow-related variables have to be calculated;
- specifying the boundaries of the geometry as sets of cell faces;
- applying appropriate boundary conditions to each set of faces;
- defining the initial conditions for the simulation.

Note that the geometrical locations within the flow, such as an inlet or a wall, have to be defined as sets of cell faces or even cells. This is because the CFD solver knows nothing of the real geometry of the problem; it only has information on the mesh of the flow domain.

This stage of the analysis process is carried out by giving commands to the pre-processor program of the CFD package. These pre-processor programs are usually interactive programs, where the commands can be entered using many of the input devices available to the user such as the keyboard and a mouse. This allows the specification of the flow to be built up in small stages. It is useful to enter the commands in groups that relate to one particular part of the specification. For example, these groups might be commands specifying

112

the boundaries of the domain or the numerical control parameters. Examples of the ways that this happens in practice are given in Chapter 10 where some simulations will be performed using commercial software. To assist the user, the status of the specification can be checked at any time by asking the pre-processor to show some part of the data.

Sometimes when entering the data for a series of similar fluid flow problems, interactive input can become a boring and repetitive process and so most of the CFD pre-processors allow a user to create a datafile with a text editor. This datafile contains the necessary input for the pre-processor. Some programs will even write such a file from the data that has already been entered and this is extremely useful as the file of commands should be error-free. Using such files of commands can save a large amount of data preparation time.

7.1 Specifying fluid properties

Fluids possess a variety of properties, as we saw in Chapter 2, and the solver program must be given some way of calculating the values of these. When solving problems with CFD two of the most important properties are the density and laminar viscosity of the fluid.

For simple problems, where the fluid is assumed to be laminar and incompressible with no heat transfer effects, the density and viscosity are taken to be constants. These constants are given to the software by simply entering the appropriate value. One possible mistake is to confuse the two ways of stating a fluid's viscosity. The standard viscosity μ, also known as the dynamic viscosity, is the constant that links the physical shear of a fluid to the shear stress, and the kinematic viscosity ν is the ratio of viscosity to density, ρ, i.e.

$$\nu = \frac{\mu}{\rho} \qquad (7.1)$$

For air, where the density is about 1 kg/m^3, any mistake is unlikely to be found from the results, but for liquids like water, where the density is 1000 kg/m^3, the result of a mistake could lead to large errors in the calculated solution. As a check, the units of viscosity are kg/m s and the units of kinematic viscosity are m^2/s. For the common fluids, tables of the density and viscosity values have been drawn up (Kaye and Laby, 1983; Haywood, 1976).

If the flow is known to have significant variations of temperature,

113

perhaps due to heat transfer, then the viscosity will vary as a function of the temperature. The pre-processor may well allow the user to specify the relationship or, at least, allow the user to switch on some standard variation of viscosity. A common variation that is used is a power-law form (Schlichting, 1979):

$$\frac{\mu}{\mu_\infty} = \left(\frac{T}{T_\infty}\right)^\omega \tag{7.2}$$

which can be seen to be a non-dimensional relationship. Here the subscript refers to a reference value of viscosity or temperature and ω is a constant which has the value of 0.76 for air. Also the Sutherland formula can be used (Schlichting, 1979):

$$\frac{\mu}{\mu_\infty} = \left(\frac{T}{T_\infty}\right)^{3/2} \frac{T_\infty + S}{T + S} \tag{7.3}$$

where S is a constant and has a value of 110 K for air.

The same is also true of density, where various gas laws can be used to find the density from the pressure and temperature of a gas (Rogers and Mayhew, 1980). For example, we could use

$$\frac{p}{\rho^\gamma} = k \tag{7.4}$$

which is the isentropic relationship for processes which are reversible and adiabatic (where γ is the ratio of the specific heats and k is a constant), or

$$\frac{p}{\rho} = RT \tag{7.5}$$

which is the ideal gas relationship, where T is the gas temperature and R is the gas constant.

When the viscosity and density vary, the flow problem is more complex than that of a simple, incompressible viscous flow. Some discussion of how these problems are solved using CFD is given in Chapter 11.

Finally, other properties may have to be defined, but which properties these are will depend on the problem. Some examples of these additional properties are the thermal conductivity of a fluid which is needed if we are simulating heat transfer problems, or an effective turbulent viscosity which is needed for the simplest turbulence models.

114

7.2 Determining the variables that need to be calculated

Once the fluid properties have been defined we need to determine which variables are to be calculated. Which variables are needed depends on the way in which the governing equations have been discretized and the algorithm set up to solve them. With the standard SIMPLE-like algorithms, the pressure has to be calculated together with some of the velocity components. In one dimension, only a single component, say u, has to be found, whereas in two or three dimensions u and v or u, v and w have to be found respectively. When we discussed the governing equations in Chapter 2, we saw that these variables completely define a laminar, incompressible flow, and could define a turbulent incompressible flow, if only we had the computer power to solve all the equations with sufficient time resolution.

As, usually, there is not sufficient computer power available to resolve the effects of turbulence, these effects have to be modelled. This means that a set of variables that are part of the turbulence model has to be calculated. Exactly which variables are required depends on the turbulence model that is to be used, and some of the models were reviewed briefly in Chapter 2. The simplest turbulence model is to specify a single value of the additional viscosity μ_T due to turbulence. This quantity can be regarded, in effect, as a property of the fluid, and its specification has already been mentioned. Other common ways of calculating the additional viscosity due to turbulence are as follows:

- *To find it from a mixing length, which has to be specified for a boundary layer or wake.* When using this turbulence model, no additional partial differential equations have to be solved but the pre-processor has to be used to give the solver some way of calculating the mixing length and an expression for converting this to the additional viscosity. This model is normally only used for very simple geometries as this makes it easy to specify the mixing length in terms of the geometry.
- *To find it from a set of auxiliary partial differential equations where one, two or even more equations are required.* The industry-standard method is the two-equation model that uses turbulence kinetic energy k and the rate of dissipation of k, denoted by ε. Despite the fact that this turbulence model can produce poor flow predictions in some circumstances, it

115

still gives usable results for many flow situations. As we saw in Chapter 2 there are other relationships that could be used, and these include algebraic stress models and Reynolds stress models.

For problems that involve heat transfer the fluid temperature, or perhaps the fluid enthalpy, must be calculated. The equations to do this are similar to those for momentum transfer. For example, in equation (2.8) the variation of the scalar variable, the velocity component u, is described, and the other variables can be treated as just additional scalar variables. Chapter 11 looks at how the effects of heat transfer are modelled and also reviews some other flow types such as compressible flow. In both of these cases the density can vary throughout the flow field and so the fluid density might be an additional variable that needs to be calculated. Equally, as the flow types become more complex so other variables will need to be calculated.

7.3 Finding the boundaries

To calculate the required variables, the governing partial differential equations must be solved and so the boundary conditions for each equation must be specified. When the flow specification was produced the boundaries were defined in terms of the geometry of the flow domain, and now we must find these boundaries in terms of the mesh that is being used. This involves defining the boundaries as a collection of cell or element faces.

7.3.1 *Boundaries for meshes with a regular structure*

If the mesh has a regular structure, a knowledge of the local coordinate system (see Section 6.3.2) can be used to define a set of indices i, j, k. These indices denote the position of a cell within the mesh structure and range from unity to the maximum number of cells in each of the local coordinate directions. The local coordinate system can also be used to define the faces of a cell within the mesh. Looking at Figure 7.1 we can see a mesh with a regular structure shown in terms of its local coordinate system. Each cell of the mesh has six faces and a typical cell is shown with its faces labelled with the points of a compass. Hence the faces are named North (N), South (S), West (W), East (E), Top (T) and Bottom (B). The first four names are fairly standard, being used by a wide range of CFD programs, but

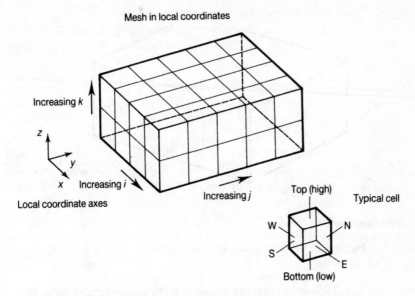

Figure 7.1. Use of local coordinate system

the last two are also known as High (H) and Low (L) in some programs. For example, the face of the cell that is at the most positive local x-direction position, in the direction of increasing the index i, is the East face and the one at the most negative local x-direction position is the West face.

We can also see, by looking at Figure 7.2, that any plane of cells will have a constant value of either i, j or k, and that the extent of the plane can be defined by knowing the limits of the other two indices. The patch of cells shown in Figure 7.2 has a constant value of the index i and the limits are defined by j_{min}, j_{max}, k_{min} and k_{max}. Also, the faces of the cells in the patch shown are in the positive local x-direction and so they are all East faces. By using this notation a set of patches can be defined on the boundaries of the mesh. These patches have to be defined for all the surfaces where the boundary conditions are not automatically specified by the solver program.

It is worth remembering that when defining a patch of cell faces on a boundary, it is sensible to define patches that will have only one boundary condition type applied on the patch for each partial differential equation. This means that the whole patch might be an inlet or an outlet, but not both. By doing this it is simple to specify the boundary condition that applies on a patch by a single command.

117

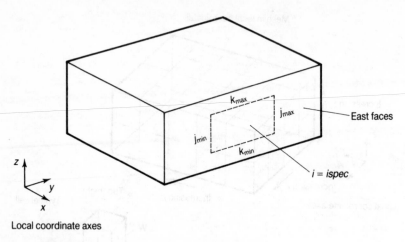

Local coordinate axes

Figure 7.2. Defining a patch of cell faces

7.3.2 *Boundaries for meshes with an irregular structure*

When a mesh has an irregular structure the problem of defining the boundaries becomes much more difficult. Actually finding the cell faces that are the boundaries of the mesh is quite straightforward, as we shall see. It is the collecting of the various cell faces into groups that are suitable for the addition of the same boundary condition that is difficult.

Two pieces of information help us to find the cell faces that are on the boundary of the mesh. First, each face of a cell is uniquely defined by the nodes that are on the face and, second, the faces on the boundary of the mesh can only be associated with one cell, whilst those internal to the mesh must be associated with two or more cells. This is shown in Figure 7.3, where it is clear that the internal face is common to the two cells and that the external faces are only related to one of the two cells.

The process of finding the faces that are on the boundary of a mesh is called a *free-face check*. The algorithm used to do this is shown in Figure 7.4, from which it can be seen that each cell is considered in turn. Then each face within a cell is found in terms of the numbers of the nodes attached to it. A unique label for each face on the cell is then found from these node numbers. Each of these face labels is then checked against a list of the face labels stored in a database. This database is created as the process is carried out and records the number of cells that a given face is attached to. If a face label does

118

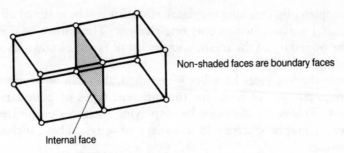

Non-shaded faces are boundary faces

Internal face

Figure 7.3. Internal and boundary faces

not exist in the database then an entry recording the new face label is made in the database and the count of occurrences of the face set to unity. If the face has been listed before, the count is increased so that it reflects the number of elements associated with the particular face. Once all the faces on a cell have been processed then a new cell is chosen, and after all the cells have been processed the database will

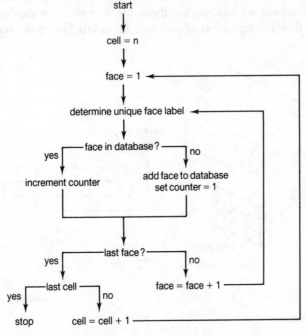

Figure 7.4. Algorithm for determination of free-faces

be complete. By checking the database, a list can be made of all those faces that are attached to only one element. These must be the faces on the boundary of the mesh, and the list of faces is known as a *free-face list*.

Once the free-faces have been identified, they can be grouped into the required sets of faces for the different types of boundary conditions. This is usually done by displaying the faces in the free-face list on a graphics screen in a variety of ways. These include the following:

- a hidden-line display, where the user sees the faces just as they would be seen if they existed physically; that is, faces that are behind other faces, as seen by the viewer, are hidden from view;
- a display of the faces within a given volume.

Once the displays of the bounding faces of the mesh have been produced the pointing device of the terminal or workstation can be used to pick out the faces. This can be done either face by face, or whole sets of faces can be picked by placing a window on the screen and noting the faces that are within the window. This is illustrated in Figure 7.5 where we can see a simple mesh with an irregular structure. The flow inlet consists of the nine faces labelled in the left hand

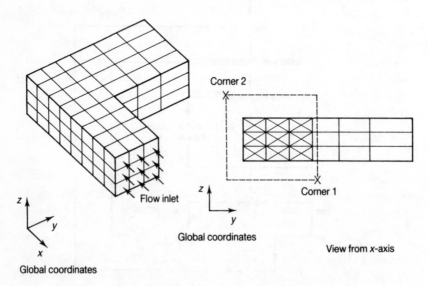

Figure 7.5. Finding boundary faces on the screen

view. These faces could be picked manually using the cursor on the display screen, but, by changing the view of the mesh to that shown on the right hand side of the diagram, a rectangular window can be defined using two corner points as shown. Then all the faces that are wholly within the window, the nine required faces, can be labelled by the pre-processor as being boundary faces. This windowing method has great advantages when dealing with large numbers of boundary faces.

7.3.3 *Grouping faces together*

Regardless of whether the mesh has a regular or an irregular structure, the boundary faces must be grouped together into sets of faces using the methods we have just described. Each set of faces can then be given an index that will allow the set to be related to a boundary condition. Sometimes, the boundary condition on a set of faces will be unique to that set; however, in some cases, the same boundary condition may well be applied to several sets of faces. In this latter case, each of the sets can be given the same index and then the index can be linked to the given boundary condition.

Finally, it is useful to know that some CFD solvers will find all the cell faces on the boundary of the mesh. This list of faces can then be compared to the boundary faces that have been specified by the user. It is common for any unspecified boundary faces to be assumed to be solid walls. This can save a great deal of effort for the user if the mesh is for a flow problem such as a complex internal flow. These meshes can have multiply connected passages, the boundary faces of which can be very difficult to view. By considering all unspecified faces to be solid walls, the user does not have to specify these faces and the saving in effort is large if this is done.

7.4 Defining the boundary conditions

Now that we know where the boundaries of the mesh are, in terms of the cell faces, and now that we have grouped them appropriately, we have to consider which boundary conditions should be applied. For each partial differential equation that has to be solved, the numerical method that is used determines which boundary conditions can be specified. In some cases one particular boundary condition must be specified, such as the specification of the velocity or pressure. In other cases, certain conditions at a boundary will happen naturally if nothing is specified there. Often, the software will predict a flow which has the derivatives of the velocity normal to a boundary calculated as zero if no other specification is made. Of course, if the analyst wishes, such conditions at the boundary can be changed by specifying the appropriate values.

In most CFD problems several different types of boundary condition are usually applied. Boundary conditions were discussed in Section 3.5.4, but a summary of the possible types is given here for completeness. When using a SIMPLE-like algorithm the common boundary conditions that are applied come from the following:

- *The momentum equations*, where the velocity components can be specified on a boundary. If this is not done then the derivatives of the velocity components in a direction normal to the boundary will be automatically set to zero. As this is the required condition at a plane of symmetry, and is often the required condition at an outlet, this automatic specification is extremely useful.

- *The pressure correction equation.* This requires that the mass flow through a boundary is specified or it will be given as zero; and it also requires that the pressure is specified at some point in the flow domain. This latter requirement comes from the nature of the pressure correction equation, which can only relate the derivatives of the pressure, not absolute values of pressure. Consequently, if the value of pressure is not specified at some point then the pressure solution is singular and cannot be found. A further complication in specifying the pressure is that at places where it is specified the continuity equation does not hold. This comes about because the continuity equation is not enforced at a point if the pressure is fixed at the point, as the specification of the pressure overwrites any information about the continuity equation at that point. If the continuity of the flow is not strictly enforced then fluid can leak into or out of a system through a point where the pressure is specified.

Whilst these are the main boundary conditions that come from the partial differential equations, flow problems are often not described in such terms. For example, during the initial specification of the problem, discussed in Chapter 5, we might have decided that the boundaries should show the following characteristics:

- *A solid wall with a turbulent flow over it.* To model this accurately requires many points through the boundary layer as the shear at a solid wall in a turbulent flow is much greater than that for a laminar flow. The computational effort required to do this can be reduced by assuming that the flow velocity varies in a logarithmic fashion through the boundary layer, as found in experiments. This was discussed in Chapter 3. Then empirical approximations to the values for the velocity at points just away from the wall can be used. Similarly, the boundary conditions for the additional turbulence parameters, such as the turbulent kinetic energy and its rate of dissipation, can be set in an automatic way to some empirically derived values.

- *A free surface.* Here, the fluid pressure is fixed but the fluid velocity and the shape of the boundary are not known. These surfaces occur, typically, when we model the surface of a liquid in contact with air, for example when calculating the flow around a ship. Special CFD programs can handle these boundaries, but if the surface shape does not need to have the effects of waves modelled, then we can use a symmetry plane as an approximate model of these boundaries.

- *Moving walls*, such as a piston in an internal combustion engine, where a solid surface moves in the flow.

- *An inlet with a turbulent flow coming through it.* Here the turbulence parameters are convected into the fluid flow domain and the levels of the variables that are brought in must be specified.

Some examples of the application of boundary conditions will be given when we look at the case studies in Chapter 10. There we will see that it is usual for the common boundary condition types to be pre-programmed options of the software.

At all the boundaries, it is possible that a given boundary condition may apply for only a fixed amount of time. This could be the case if the problem is time-dependent, for example when modelling the opening or closing of a valve. In these cases, for each patch of cells or each boundary index, the CFD pre-processor can be used to assign the appropriate boundary condition and the duration of its application.

7.5 Defining the initial conditions

Many solution algorithms require that some form of initial flow field is specified for the solver. This could be due to the flow actually being time-dependent, where the initial state of the variables is required to start the calculation, or it could be due to the CFD solution algorithm using a quasi-time-varying solution algorithm to calculate a steady state solution. Equally, the non-linearity of the problem will demand some initial guess for the variables which will have to be supplied either as a series of default values or by the user. Chapter 3 discusses these factors.

In all cases any initial flow field must be specified for every cell in the flow domain. Usually, the specification of the initial conditions is fairly straightforward, as some simple flow field can be given such as the flow being at rest with zero pressure everywhere or some uniform fluid motion could be specified such as that calculated from a potential flow solution. Such a solution is an ideal flow solution which would occur if a fluid had no viscosity and could not be compressed. Sometimes the initial conditions are specified for groups of cells with a constant value of a variable being set within each group.

If turbulence variables such as k and ε are being used, then they are usually set to a small positive value or to some realistic value. This is done to prevent an error occurring during the calculation procedure where the program attempts to divide by zero when these variables are being used. Ways of calculating the size of the initial magnitude of these variables will be discussed in Chapter 10 when some simulations are performed.

7.6 Using user-generated sub-routines to influence the simulation

Each pre-processor will allow the user to specify the properties, boundary and initial conditions for a wide variety of flow problems. This is usually sufficient for most CFD simulations that will be calculated. However, there will always be an exception to this and sometimes the user will want to define some information that is not standard. To allow this, some CFD software systems allow users to write their own computer programs which can influence the workings of the solver.

One common way of doing this is for the user to write some FORTRAN sub-routines that are linked into the solver program, or lines of FORTRAN code can be written into some general access sub-routine that is provided by the CFD software supplier. This sub-routine is then compiled and the object code linked with the main solver program libraries to provide a new, modified solver program.

This might sound straightforward but in reality it is very difficult, as the user has to find out so much about the way the solver has been written. Users should be extremely familiar with CFD simulations before they embark on writing their own software and embedding it into the solver. This is a technique for the expert in the use of CFD.

8 Obtaining a solution

As we have now created a mesh to describe the geometry of the flow domain (see Chapter 6), and also specified the properties of the fluid and the boundary and initial conditions of the problem (see Chapter 7), the actual flow problem is completely defined. This means that the CFD software should have all the information that it requires about the flow. We are nearly ready to run the solver program and obtain a solution.

This chapter looks at the final preparation of the data and the running of the solver. In particular we will discuss the following:

- *How to set up the data for the solver.* As the simulation is achieved by a numerical transformation of the governing equations, we must specify the information that is required to control the numerical solution algorithm. Further, administrative information such as the form of the output of the solver program must be specified.
- *Running the solver and then analysing the output to identify any problems that have occurred.* These can then be rectified before running the solver again to obtain a better solution.

We mentioned in Section 4.1 that the whole analysis process will not, normally, be carried out just by executing the list of tasks one after the other. Sometimes we must run the solver, check the results and then rebuild the computer model so that the simulation is improved. Often, the production of a good simulation will be a continuing process of trial and error.

8.1 Final data preparation

8.1.1 *A note on iterative processes*

When using a CFD package the details of the numerical solution process will usually be hidden from the user. However, some features of the process are common to all packages and the controls that have to be used are often similar, even though the values of the control parameters may be algorithm- or problem-specific or both. In particular, as we saw in Chapter 3, the non-linearity of the equations forces the solution process to be iterative, regardless of whether the problem is time-dependent or not. This means that an initial solution, normally a guessed solution, is required at the start of the solution process, and then the numerical equations are used to produce a more accurate approximation to the *numerically correct* solution, which is one in which all the variables satisfy the governing equations. This new approximation, the updated solution, is then used as the new starting solution and the process is repeated until the error in the solution is sufficiently small. Each repetition of the solution process is known as an *iteration*.

Sometimes during an iterative process the updated solution at the end of one iteration can be very different from the solution at the start of the iteration. If we consider Figure 8.1 we can see a graph of velocity against time. Let us imagine that we have a numerical scheme that predicts the velocity V_{new} at some time Δt ahead of the current time by using values of the current acceleration a and the current velocity V_{old} in the following way:

$$\frac{V_{new} - V_{old}}{\Delta t} = a \tag{8.1}$$

or

$$V_{new} = V_{old} + a\, \Delta t \tag{8.2}$$

which is a first-order method in time. If we know both the current acceleration and velocity then we can predict the new velocity, and so, given the new acceleration and velocity, we can march forward in time finding the velocity–time relationship. Looking at the diagram we can see the actual velocity–time relationship and two approximations based on the above equations. In both of these the initial acceleration is used to predict the velocity. It is clear from this that if the time interval is small, say Δt_1, then the error ε_1 between the predicted velocity and the actual velocity is small, but if the time

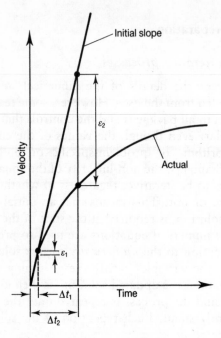

Figure 8.1. Errors in prediction

interval Δt_2 is large then the error ε_2 is large. Similar errors can occur when carrying out a CFD simulation and if the error gets ever larger during the solution we will have a very inaccurate flow solution and convergence of the solution will not be achieved (see Section 3.4.1). So that we can see whether or not this is occurring we need a measure of the error of the solution.

Fortunately, the numerical equations that we wish to solve can also be used to find such a measure of error. These measures of error can be used to see if a solution process is converging, and they are known as *residual errors* or *residuals*. At the end of each iteration, the latest solution can be used to generate all the terms in the various partial differential equations. For example, if all the terms in the momentum equation, equation (2.8), are placed on the left hand side of the equation and the individual components of the equation are formed from the solution for the velocities and pressure, then these terms can be summed and the sum should be zero. As the solution is only an approximation to the required values of the variables, the sum will not be zero. It is this sum that is the residual error.

As the solution process progresses from iteration to iteration, the

128

residual errors from each equation should reduce. If they do reduce then the solution is said to be 'converging'. If the residuals become ever larger, then the process is said to be 'diverging'. Most CFD solvers write the residuals to a datafile, or even to the terminal screen, at the end of each iteration. This enables a quick check to be made on the progress of the solution.

If the solution scheme is time-dependent or quasi-time-dependent, then the solution at the end of each time step needs to be converged before moving to the next time step. This can mean controlling several iteration procedures. As we saw in Section 3.5.1, one iterative procedure might solve the simultaneous equations generated by linearizing the partial differential equations; the second iterative procedure finds a solution at one time step and accounts for the non-linearity of the problem; and a final iteration procedure, if required, moves the solution through the different time levels. All these iteration processes need to be controlled.

8.1.2 *Controlling the iterative processes*

To prevent the whole solution process from diverging, when the residual errors become larger instead of becoming smaller from iteration to iteration, we must control all the iterative processes in some way. In Section 3.4.2 we discussed the use of iterative solution algorithms to provide solutions to a set of simultaneous equations. If these are used by the CFD solver program, then the controls are often built into the program, but occasionally it may be necessary to provide values for the number of iterations that are to be performed in solving the simultaneous equations, as well as values for the relaxation factors. In CFD calculations it is always important to ensure that the velocity field, used in the momentum equations, satisfies the continuity equation. This means that when using a SIMPLE-like algorithm more iterations are used to solve the pressure correction simultaneous equations than are used to solve those from the momentum equations.

Turning to the control of the other iterative procedures, two methods are commonly used. For steady state problems the terms in the equations which contain the time variation are often left out, and so the solution generated by this type of algorithm has to be controlled by using relaxation parameters. These take the solution calculated during the current iteration and scale it so that the solution used in the next iteration is not too different from the solution at the start of the current iteration. This is done by using a relaxation factor ω, and the scaling of a variable ϕ can be calculated from

$$\phi_{new} = \omega\phi_{calc} + (1 - \omega)\phi_{old} \tag{8.3}$$

Here, ϕ_{old} is the value of a variable at the start of the current iteration and ϕ_{calc} is the value of the same variable calculated at the end of the iteration. The relaxation process given in equation (8.3) uses these two values of ϕ to produce a value of ϕ, i.e. ϕ_{new}, which is between ϕ_{calc} and ϕ_{old}. The solution ϕ_{new} then becomes ϕ_{old} for the next iteration. This scaling uses values of ω which are between zero and unity and is known as *under-relaxation*. Note that if ω is unity there is no relaxation and that if ω is zero then the solution does not change at all. Intermediate values of ω provide scaling between these extremes and enable the user to prevent divergence of the solution process. Looking at Figure 8.1 again, a reduced value of velocity obtained by iteration and relaxation would be more accurate if the time step was too large. Note that the scaling is carried out for every value of a given variable, that is, at each node or cell.

In CFD simulations where relaxation factors are required to control the overall iteration process, the factors are usually applied to all the variables, with ω normally being set in the range 0.1 to 0.3 for the pressure solution and in the range 0.5 to 0.9 for the velocity solutions. If the $k-\varepsilon$ turbulence model is used, then the ω-values for these two equations are set to be the same as those used for the velocity solutions, or to lower values. If a mesh is complex and the cells are not near-cuboid in their shape then the relaxation factor applied to the turbulence variables might have to be much smaller, say up to ten times smaller, than the relaxation factor applied to the velocity variables.

The second means of controlling the overall solution process is to use a time-dependent solution scheme, even if the flow is known to be steady. Such a scheme mimics the physical changes that a flow would undergo if it were changing with time, as the modelling of the time variation smooths out the way in which the solution changes from one iteration to the next. With time-dependent schemes the main controlling factor is the value of the time step. This is set to give as small a number of time steps as possible whilst maintaining a smoothly converging solution. For steady state problems, only the converged solution, after what is effectively an infinite period of time, is required and so the time step can be large; but for transient problems, when the time variation is of interest, the time step must be small enough to model the temporal changes in the flow variables accurately.

It is difficult to give specific rules for calculating a value of the time step that will always give a converging solution, as the stability criteria of the Navier–Stokes equations cannot be found analytically. A time step of the order of the *residence time* t_{res} of a fluid particle in a cell is often used. This is the time it would take a fluid particle to move through a cell. For example, if a fluid particle moves in the x-direction, the residence time is given by:

$$t_{res} = \frac{\Delta x}{U} \tag{8.4}$$

where Δx is the length of the cell in the x-direction and U is the fluid velocity in the x-direction. These values are found for some typical cell in the flow field. This works well for the momentum equations which calculate the velocity components, but the time step may have to be reduced by a factor of, say, 100 for the other transport equations such as those for the turbulent kinetic energy k and its rate of dissipation ε when the standard two-equation turbulence model is used.

8.1.3 *Other solution control information*

Having decided how to control the iteration processes that take place, we can now use the pre-processor to build up the remaining information that is required by the solver. As well as the iteration control information that includes the relaxation and time step parameters, we must give the solver some or all of the following:

- *The number of time steps to run.* This will be one step if the solver is to produce a steady state calculation.
- *The number of iterations to carry out*, within each time step, whilst resolving the non-linearity of the problem.
- *The number of internal iterations required* in solving the simultaneous equations (if iterative methods are used to do this).
- *Limits on the residual errors.* Using these limits prevents computing effort being wasted in trying to compute the solution to some ridiculous numerical accuracy. Once all the residuals fall below this limiting value the calculations are stopped.

131

- *The form of the discretization of the convection operator in the momentum equations.* Various methods were discussed in Section 3.5.3. For CFD calculations that involve complex geometry it is best to start the calculation with a discretization which will be likely to produce a converging solution. This can often mean that the solution will be inaccurate due to the diffuse nature of the discretization, as was explained in Section 3.5.3.

- *The data that the solver should store in files or write to the screen.* This data should include all the values of the variables that are calculated so that they can be analysed with the post-processing program and also read again by the solver program if the calculation has to be continued. If the solution is time-dependent several sets of solutions at various times might be required. Also, we will want to check the residuals of all the variables and so these are written to a file. As a further check on the convergence of the solution, most programs allow the user to specify a location in the mesh, say one cell or node, at which the program will write the values of the variables at each iteration or time step.

- *The destination of the data that is to be produced.* Some of this data will be written to datafiles, some will go to the screen. The location of the files, perhaps a directory on a disk of the computer, will need to be known by the software.

Once these choices have been made, the specification process is completed by entering the values using the pre-processor. In some cases the CFD software as supplied will not write or even calculate some of the required data itself. In these cases the user has to write computer program sub-routines that can be linked into the solver. These sub-routines are used to produce the required data from the information stored by the solver, and some ways of doing this were mentioned in Section 7.6. It should be remembered that this is the realm of the CFD expert.

When all the data has been prepared, the pre-processor can be instructed to write the datafile or files that will be accessed by the solver program and the solver can be run.

8.2 Running the solver and troubleshooting

The way in which the solver program is run will vary from package to package. It is common, however, for a small set of computer operating system commands to be written that will automate the process. This can be done by either the user or the supplier of the software. These commands make sure that the correct datafiles are accessed, possibly copying them to another machine if the solver and pre- and post-processors run on different machines. They also run the solver program and then they return the results files to the user-specified location if this is required.

At the start of any analysis, the user should instruct the solver program to perform only a few iterations. This enables the user to perform convergence checks on the solution process by looking at the values of the residuals either on the screen or in a datafile and seeing if they are reducing or increasing. After running something like ten iterations the initial trends in the residuals should be clear. If they are reducing the solution process is clearly converging and this is the desired situation, whereas if they are increasing, further thought is required before the convergence properties of the solution can be determined. Some typical graphs of the residual value for one of the flow equations plotted against iteration number are shown in Figure 8.2.

Often, there is a large increase in the residual value in the first two or three iterations, but this is nothing to worry about if the residuals fall after this, as shown in Figure 8.2(a). However, if the residuals are still increasing after ten iterations then the differences in the residuals from iteration to iteration need to be examined. If the difference is increasing from iteration to iteration, the process is diverging (Figure 8.2(b)), but if the differences are reducing then the process is

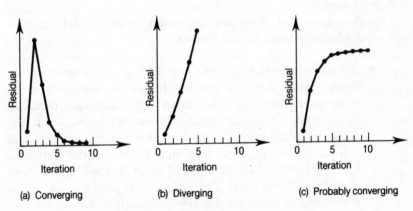

Figure 8.2. Initial trends in residual variation

probably converging (Figure 8.2(c)) and it is likely that the residuals will start to reduce in value if the solver is run for more iterations.

When the process is seen to be converging, then the pre-processor can be used to increase the number of iterations, to, say, 100, and the simulation continued. At this point the solver must also be told to use the last solution that was calculated as the new initial solution. Hopefully, this solution was stored in a datafile at the end of the first run of the solver program. As this prevents computer time from being wasted, do not use the initial values that were set before the solver was run. When the latest values of a solution are used as the initial solution, the calculation is known as a *restart calculation*.

If divergence occurs, then the first remedy is to check the computer model for obvious errors. This can be done by reading any of the input data that has been written by the solver program and by meticulously checking the data stored by the pre-processor. The computer model should reflect the original specification that was produced at the beginning of the analysis. If nothing obvious is found, then the next step is to change the relaxation factors or the time step. When using relaxation factors, if the value of ω for pressure is already about 0.1, then the values for the velocity and the turbulence modelling quantities must be reduced. Usually, the turbulence variables must be relaxed more than the velocity, and so the relaxation factor applied is smaller. This process of changing relaxation and time step values is very much a question of trial-and-error and so it can involve running several initial sets of, say, ten iterations, each with different relaxation values. If a converging solution cannot be achieved, then it is probable that there is some sort of error in the computer model.

Common causes of divergence related to poor modelling, and some possible solutions, are as follows:

- *A poor mesh which has cells that differ greatly from a cuboid shape*. This is a typical problem if a finite volume scheme is being used which has some terms in the numerical formulation missing. These terms might describe, for example, the non-orthogonality of the mesh as detailed in Section 6.3.2. It is these missing terms that should enable the calculation to be accurate on a non-orthogonal mesh. Smoothing the non-orthogonal mesh using a procedure that produces an orthogonal or near-orthogonal mesh might help to overcome this.
- *Inadequate prescription of the boundary conditions*, such as

not specifying the pressure anywhere. This has to be dealt with by carefully checking the data defined with the pre-processor.

- *Poor initial conditions*, i.e. unrealistic and too far from the the conditions that exist if the solution is converged. One way of improving the initial conditions is to run a potential flow solution first. Such a solution assumes that the flow is both inviscid and incompressible. This, of course, will not take into account any effects of flow separation.
- *Applying insufficient upwinding for the convection terms*, as was discussed in Chapter 3. Smoothing the mesh might help, but the use of a more diffuse upwinding scheme usually cures the problem, at the expense of getting a less accurate solution than would have been hoped for.
- *The turbulence model.* Running the solver with the simplest turbulence model, i.e. just specifying an effective turbulent viscosity everywhere should enable some results to be obtained. These can then be used to start a new calculation with a more sophisticated turbulence model.

If modelling errors are found they must be corrected, either by changing the mesh or by using the pre-processor to modify the input data. Then the solver can be run again and the solution process checked all over again.

Eventually, it should be possible to achieve a converged solution. This is a solution where the residuals are several orders of magnitude lower than the maximum value recorded during the solution process. Once a converged numerical solution has been found all that we can be sure of is that *the numerical solution satisfies the numerical equations* on the mesh we have used to some order of accuracy. What we require is that the converged solution will bear at least some relationship to the physical flow that would be obtained. Usually, this is the case, but we must check that the converged solution is reasonable in the light of the expected flow structures, as discussed in Section 5.1, and as illustrated in the examples of Chapter 10. The difference between the physical flow and the numerical solution could be due to one of the following:

- an inadequate mesh density being used in regions of high rates of change of the flow variables, for example in a boundary layer;

- inadequate physical modelling of the flow, especially due to the use of turbulence models which are too simplistic; for some flows this is all turbulence models;
- poor specification of the boundary conditions which have over- or under-constrained the flow, typically at an outlet to the system where the pressure has been fixed as a constant; this restricts the flow if it swirls out through the outlet as the calculated pressure needs to be able to vary across the outlet to provide the necessary centripetal force.

Sometimes it is possible to see these errors during the post-processing phase and we shall look at various examples of this in Chapter 10 when we produce some flow simulations. Finding these errors is really a matter of experience.

9 Analysing the results

In the preceding chapter we looked at how to obtain a set of results using the solver program. These results should be a converged numerical solution to the governing equations, produced with the appropriate boundary and initial conditions on a mesh that describes the geometry of the problem. Remember that the solution is strictly a solution of the numerical problem, not of the physical problem, and that the differences between these two could be due to such things as an inadequate mesh or a poor turbulence model.

When the numerical solution is obtained it is necessary to determine whether or not it bears some relationship to the physical reality. If it is likely that it does, then the required technical information can be extracted from the results. This chapter looks at what the results of a simulation are, how computer graphics can be used to obtain pictures of the results, how the solution can be checked to see if it is likely to be reliable, and finally how the model can be refined so that the required data can be obtained from the results.

9.1 The results obtained from the solver

When the solver runs it produces a large amount of data that has to be analysed. This analysis might be undertaken so that some cause of divergence in the solution process can be identified, so that the quality of the solution can be examined or so that useful technical information can be extracted if it is a converged solution. First, we must consider what information will actually be available to us when we want to analyse the results.

Information can be produced by the solver in two main forms. These forms differ in how the data is stored by the computer. In one form the data is stored using an internationally agreed format that defines individual characters of data such as the letters of the alphabet or the numbers 0 to 9. This form of data is known as *ASCII data*, after the committee that devised the data standard, and can be written to a terminal screen or stored in a file known as an *ASCII file*. Each character has to be defined by one byte, i.e. eight bits, of computer memory, and so 256 different characters can be specified. ASCII files of data can be edited by text processors and other software, and they are effectively machine-independent, which means that the data can be transferred from one computer to another computer, even if the machines are from different manufacturers, without any translation process taking place. Whilst most computer manufacturers use the ASCII standard, there are other standards such as EBCDIC which are used by a minority of manufacturers.

Numerical data can also be stored in the second data storage format, which is known as *binary data format*. There is a standard for this method of data storage, but usually, in 1991 at least, the method of storage is peculiar to each computer operating system or computer manufacturer. Each of these binary storage methods enables real numbers, for example, to be stored by four bytes in single precision or eight bytes in double precision. Binary data is stored in files known as *binary files*. These files are not machine-independent and so cannot be transferred from computer to computer without some form of translation process taking place. Sometimes when a workstation, for example, is connected to a mini-supercomputer a translation program will be provided by the workstation vendor to facilitate the transfer process. By using binary files to store real numbers, there is a saving in the amount of storage required, as can be seen from the number of bytes required to store each number.

The type of information produced by the solver program can usually be controlled by the user but it often consists of the following:

- *Values of the residual error for the various partial differential equations that have been solved*. These are listed as a function of the iteration number or time step. As was explained in Chapter 8 these values give some idea as to whether the solution is progressing to a converged solution. This is usually stored as ASCII data so that it can easily be read later.

- *Values of some of the variables at a limited number of locations*, known as *monitor locations*, for every iteration or time step. This data also gives an indication of the progress of the solution towards a converged solution. For time-varying solutions it also gives a limited history of the development of the flow with time. Again, this is usually ASCII data.
- *A complete list of the flow variables at all the nodes of the domain or all the cells of the domain*, as appropriate, for the way in which the solver works. These lists, also known as *dumps* of the data, are produced at the end of the solution process, but the solver can also be instructed to produce such a list at intermediate stages in the process. This might be necessary if the results at several discrete times are needed to describe a time-varying flow. This is normally binary data to reduce the storage space that is required, but ASCII forms can also be requested to make reading of the data easier, if the amount of data is small, or to allow a transfer between computers.
- *Mesh data*. This is sometimes produced by the pre-processor but might be produced by the solver program. It includes the coordinates of the points in the mesh and, if necessary, the connectivity list. Depending on the CFD package, such things as cell volumes and face areas might also be stored. This data is usually held in binary form to reduce the required storage, but, again, ASCII data could be used for the same reasons as those given for the flow variable data.
- *Some form of ASCII file that reports on the progress of the solution*. This file might include an echo of the input data from the pre-processor so that the input actually used by the solver can be checked, a repeat of the residual values and monitor data at each iteration or time step, any user-programmed results, such as the pressure drop between two points or the integrated values of pressure to give a measure of the pressure-derived drag and lift on an object, as well as accounting information such as the length of time that the solver took to run and the amount of disk resources used.

In Chapter 8 we discussed how the residual errors can be analysed and a converged solution produced. Now, in this chapter, we are concerned with how the flow data at all the nodes or cells in the mesh can be analysed. Large quantities of this data are produced by a CFD solver, especially if the mesh is complex and has a large number of nodes or cells, as might be the case for an industrial flow problem. Only when small test cases are run is it possible to read the ASCII files that contain the solution, and so for realistic problems we have to resort to the use of computer graphics techniques to analyse the results visually.

9.2 Using computer graphics for CFD

9.2.1 *Using graphics hardware*

Before considering what can be done with computer graphics let us think about the hardware that is required to drive the software that will generate the pictures as well as to display the pictures themselves. A typical hardware installation will consist of the following devices:

- *A screen or visual display unit (VDU) that is able to produce a grid of points in a variety of colours.* These points are known as *pixels*, as we said in Section 4.2.2. The resolution of the screen is determined by the number of pixels that can be displayed and most graphics screens can display a grid of something like 1000 pixels in the horizontal direction by 1000 pixels in the vertical direction. If the display is monochrome then each pixel can only be shown as either black or white, whereas if the display is a colour device then each pixel can be displayed in one of several colours. Typically, sixteen colours or even 256 colours are used. The screen could be part of a terminal which is attached to a computer or it could be part of a workstation.
- *A keyboard,* which allows the user to interact with the software by typing commands and replying to questions from the software.
- *A pointing device,* which should enable a cursor to be moved around the screen. This pointing device could be a *mouse,* which is a small device that senses movement either mechanically or optically, or it could be a simple set of four direction keys.
- *A button box.* This is used in the more expensive installations to manipulate the picture. The box has several knobs

140

on it that can be used to rotate an existing picture about any of the three coordinate axes, or to zoom in and out or pan across the picture.

When the user runs the graphics software, the program should activate the screen, keyboard, pointing device and button box in such a way that the user can develop an intuitive feel for the manipulation of the results.

9.2.2 *Using graphics software*

The graphics software itself is usually supplied as part of the CFD software package and is known as a post-processor. Sometimes, however, this software is combined together with the pre-processor to form a single interactive program that is used for both creating the computer model and post-processing. Also, post-processors from other sources such as finite element structural programs might be available and these can also be used.

These programs enable a user to see the geometry of the flow problem, the mesh and the results of the simulation by producing pictures of the available data, usually in colour. Displaying the data in a visual way condenses the vast amount of information that a CFD solver can generate into a usable format. As computer power becomes cheaper, graphics software is often run on interactive colour workstations which have sufficient display resolution for the task and also have enough of their own computer power to produce detailed pictures in a reasonable time without having an impact on other users on the network.

By entering commands the analyst can use the CFD post-processing software. These commands direct the software to build up the required picture of the data on the graphics screen. Several commands may be needed to create a picture and, in many cases, the analyst will want to generate similar pictures from one analysis to the next. To prevent the user from re-entering a lengthy set of commands it is often possible for the software to read the commands from an ASCII file. This file can be created by the user with a text editor or it could be written by the software itself in some cases.

When generating the pictures, the stages that are followed are similar regardless of the type of data being displayed. The display process involves, first of all, displaying some part of the geometry or mesh on the screen. This could be a collection of the basic entities that make up the geometrical hierarchy, Figure 6.6, or the boundaries of the mesh or even some part of the mesh itself. Then the picture is manipulated so that the required view is displayed before the solution itself is shown. This final display might be some of the velocity data, shown as a set of vectors, or the contours of scalar variables such as the fluid pressure or the turbulence modelling variables. These three stages – show the geometry, modify the view and display the results – can be performed in any order but it is usual to display the actual results last of all. As this post-processing part of the analysis process is highly interactive, the user can often move between these three stages in a seemingly random fashion. However, for most simple cases, it will be most useful if the order given above is followed. The following sections deal with each of these three stages in turn.

9.2.3 *Plotting the geometry*

When the post-processing software is started it has to read the files of results and mesh data. Then the user has to find the required view. One way of doing this is to plot some part of the geometry, normally a part of the mesh used in the solution process, onto the graphics screen. This can be done by asking the program to display the basic entities used to create the mesh, if it has access to this data or the boundaries of the mesh or the mesh itself. Exactly which of these is used will depend on the capabilities of the CFD software itself and the user's preference. A simple plot of the boundaries of the mesh is usually good enough at this stage.

Once some part of the geometry has been displayed, the user can begin to manipulate the view of the geometry so that the particular section of geometry that is required to be the centre of interest is displayed on the screen. For example, in the next chapter we will produce the simulation for the flow about a car. One area of interest is the rearscreen and boot of the car where the flow separates from the vehicle surface. To plot the results of the simulation in this area, we display the outline of the car and then change the view so that only the required area is visible. Techniques for carrying out this manipulation of the view will be discussed in the next section.

Another use of the plotting of the geometry or mesh is to check that the geometry looks like the physical situation, and also to check the integrity of the mesh. By 'integrity' we mean that the mesh should both represent the required flow domain and be structured in the correct way. The display of the mesh will show a user the basic cells or elements that have been used in the calculation procedure, and so any significant errors in the mesh, or bad modelling practice, can be found.

The way in which the mesh is displayed depends on the mesh structure that is being used. If the mesh has a regular structure then the local coordinate system and the point or cell indices can be used to specify areas of the mesh, just as was done in Section 7.3.1. Sheets of cell faces can be defined in this way and then displayed. On the other hand, if an unstructured finite element or finite volume mesh is being used then the cells can be grouped in some way and the group projected onto some cutting plane in space. Another way of displaying the mesh is to draw only the free faces of the mesh. In the next chapter we will show how some of these methods can be used.

9.2.4 *Obtaining the required view*

Once the geometry has been plotted, the view of the geometry may well have to be manipulated. There are an infinite number of ways of looking at any image and so there must be some means of defining the exact view that is required. The picture on the screen is drawn as if a single eye is looking at the object being drawn. This situation leads the graphics software to require the user to define a few fundamental pieces of data. This data can include such things as:

- *The target point*, which is the point in space at which the eye is looking.
- *The eye position*, which is the point in space at which the viewing eye is placed.
- *The up-direction*, which defines where the top of the picture should be.
- *The viewing area*, which enables the apparent size of the objects in a view to be specified.

Looking at Figure 9.1, the target point is taken to be at the origin of a set of Cartesian axes. This target is shown being viewed by a single eye which can be placed in two different positions. Default values are always given by the CFD software for the initial specification of both the target point and the eye position. These could be something like the origin and a point on the x-axis, respectively, such as eye position 1. When plotting data that relates to engineering work, the eye will normally be at an infinite distance from the target and so the effects of perspective are not seen. This means that even

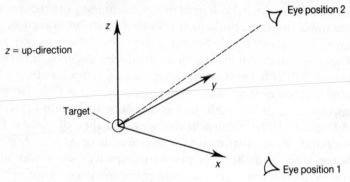

Figure 9.1. Target and eye positions

though the eye position can be defined as a point in space the software will actually place the eye at infinity on the same directional vector that joins the eye position and the target position. So, it can be seen that it is the combination of the eye position and the target point that defines the vector along which the eye looks. For some work, however, such as architectural drawing or aesthetic design, perspective effects can be produced by the software and then the eye position will be the actual point in space at which the eye is placed.

Defining these two positions in space is still not sufficient to specify the view of an object. Humans have a sophisticated balance system and this gives us information as to which is the vertical direction and so where *up* and *down* are. Computers are not as sophisticated and so they have to be told where the vertical direction is. This direction is also known as the *up-direction*. In Figure 9.1, the up-direction is in the positive z-direction. A simple example of how the up-direction is used can be seen by considering the example of the flow about a car again. We know that a car roof should be the furthest from the ground and so the up-direction will be from the ground to the roof. The vector definition of this direction, within our computer model, will depend upon the orientation of the mesh and so upon the way the mesh was built. For example, it might be in the positive or negative global z-direction, or the positive or negative global y-direction, or the positive or negative global x-direction, or any one of a host of other directions. Consequently, we must tell the post-processor which direction the up-direction is, if the pictures that it produces are to have the car in a realistic orientation. One command usually enables this direction to be specified and some examples of the effect of the command are given in Figure 9.2.

If the up-direction cannot be specified to the post-processor, as is sometimes the case, then the picture has to be orientated by a series of rotations about the three coordinate axes. This is usually achieved by specifying the angles for each global coordinate axis, x, y and z, through which the axes are to be rotated. It is difficult to produce the correct view this way using a single command. Several attempts may be needed to get the picture right.

Once the eye position, target point and the picture orientation are known, the display software can take the three-dimensional data for the geometry or mesh and draw it on the screen, in what is of course a two-dimensional representation. This can be done in one of two ways. The original way that this was done was to transform the three-dimensional data into two-dimensional data using the post-processing software. This two-dimensional data can then be plotted. Many systems still use this technique, but a more recent way of handling the data is for the post-processing software to send the three-dimensional data to the display hardware, together with the current eye position, target point and the vertical orientation. The transformation of the data from this set of three-dimensional vectors into a two-dimensional picture is then carried out within the hardware itself by a combination of both hardware and software, known as *firmware*. This local transformation is extremely fast as the firmware is dedicated to the task. Once the three-dimensional data is stored by the firmware it can be manipulated into further pictures very easily and quickly, and this is where the button box, mentioned in Section 9.2.1, can be used very effectively to modify the target point, eye position or orientation, signalling the firmware to produce the new pictures so fast that the objects can be moved in real time.

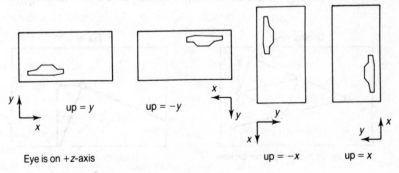

Eye is on +z-axis

up = y   up = −y   up = −x   up = x

Figure 9.2. Setting the 'up'-direction

145

Original view   New view

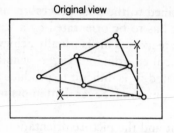 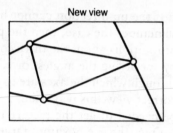

X Cursor-picked points

Figure 9.3. Zooming-in

Quite often, we wish to focus our attention on one particular area of the model, for example to see the detailed flow around a corner of an object. This can be done by changing the target position and the viewing area. The mechanics of doing this with the post-processor can vary, but there is nearly always a *zoom* command or a *centre* command. Figure 9.3 shows an example of the zoom command being used. This allows a rectangular window to be placed over the current view by defining the two ends of one of the diagonals of the window with the cursor. The software then modifies the target position and the view area to display the picture within the limits of the window. This is done whilst ensuring that the aspect ratio of the geometry is preserved. The centre command works in a similar way, but the user has to define the required centre of the new view, together with the magnification required, as shown in Figure 9.4. By using these commands in the correct combination, the view of a mesh or the results can be infinitely varied.

When working with very complex meshes, and the associated results, the sheer volume of information displayed can be too great.

Original view   New view

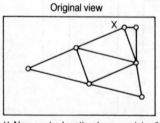 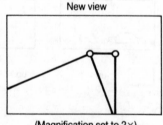

X New centre location (cursor-picked)   (Magnification set to 2×)

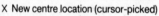

Figure 9.4. Centering and magnifying a view

146

The information content can be restricted by using the following techniques:

- *Volume clipping*, which enables the user to give limits in the global coordinates x, y and z within which objects are displayed, but outside of which they are ignored.
- *Suppression of hidden lines*, which calculates whether something that would be drawn is hidden from view by any other object, such as, for example, a cell face. If the object is hidden from view it is not drawn. The displays that are generated using this method are often called 'hidden-line' displays.

9.2.5 Displaying the results

Now we have looked at how the geometry or mesh of the model can be displayed and we know how to orientate the view to give the desired picture. Once this has been done we can add some of the results to the picture. The results that can be viewed graphically have to be derived from the flow velocity data or from the scalar data for quantities such as the fluid pressure and the turbulence variables. This data is known at a series of points in space which might, for example, be the nodes of a mesh or the centroids of the cells.

With a mesh that has a regular structure the results data can be drawn for a sheet of cells or nodes, in the same way as the mesh can be drawn. It is worth remembering that this sheet may not be planar in global coordinate space, as even a mesh with a regular structure can be curved in space so that it fits around an object. When the mesh has an irregular structure the display of results is not so straightforward. As there is no simple way of referring to a group of cells, many post-processors allow the user to define a geometrical plane through the mesh onto which the results are interpolated. This plane is known as a *cutting plane*. Other ways of grouping cells can also be used, such as showing a hidden-line plot of the results which displays only those results on the boundaries of the mesh, or displaying the results for a restricted group of cells defined by creating a list of cell numbers.

No matter which way is used to display the data, there are essentially the following two types of results display:

- *Vector plots*, which show the vectors relating to the velocity results.

147

- *Contour plots*, which show contours of the scalar variables over the domain.

Dealing with vector plots first, the vectors are displayed within the picture as arrows in two dimensions. These plots are what we see when the so-called wind arrows are shown on weather forecasts. Plotting velocity information in this way can lead to confusing displays being produced as information is lost. The arrows that are drawn are the projections of a three-dimensional vector into two dimensions. Take, for example, a vector pointing directly out of the page; this would be displayed as a point. So that some of the lost information can be retrieved, the arrows are often colour coded to denote the absolute magnitude of the vector which is the local flow speed. Usually, red denotes a high speed and blue a slow speed with intermediate shades denoting the speeds in between. This does not work terribly well on monochrome terminals!

One other problem that has to be dealt with concerns the length of a typical vector arrow. Depending on the problem, the user will want the length of the arrows to give as informative a display as possible. This means that the user must scale the arrows appropriately, either by letting the computer draw some arrows and then scaling them, or by giving the computer a typical velocity which might represent, say, 10 per cent of the screen width.

For meshes which have very dense cell distributions the arrows may be so close together that too much is displayed and the useful information is obliterated. This can be overcome by the software interpolating the velocity data onto a coarse, regular grid of points. The user specifies the distance between the points in the grid, and the arrows are drawn at the points. One problem with this type of display is that the true nature of the computed velocity field can be hidden from the user. Sometimes it is better to display the data at the position that was calculated, and we shall see why this is so when we look at some real data in Chapter 10.

Turning to contour plots, these are pictures of the lines of equal scalar value of some variable plotted through the domain. They are similar to the isobars we see on maps for weather forecasts. Little interaction is required to produce these plots, except perhaps to specify the number of contours that are to be drawn. Typically, about ten contours will be calculated, and again these will be colour coded in the picture to show the value of the variable on the contour.

A coding scheme which is similar to that used for the magnitude of a vector is used in this case as well. Sometimes, the contour levels can be chosen by the user to give the required values. This is done where several separate pictures of contours have to be produced to create the required display, and it provides a consistent display.

A variation of the contour plot is to use a surface plot. This is generated by displaying a three-dimensional surface, the height of which above a plane is a measure of some variable. This variable should be a function of the two dimensions that describe the plane. Effectively, the display shows a series of mountains and valleys.

9.2.6 *Special displays*

All of the above is applicable to the production of two-dimensional images of the data at a given point in time. Sometimes, such representations may not convey enough information to a user. One such situation is when the data describes a time-varying situation such as the flow of air into a combustion chamber of a four-stroke internal combustion engine. To provide a better feel for the results, animation can be a useful display technique. If several sets of results for, say, a scalar variable such as pressure can be stored by the solver, specialist software can read the data together with the variation in time of the physical geometry and produce a series of pictures at various times on the correct geometry for the time concerned. These pictures can be seen as the *frames* of a moving picture and the display software can be used to show these pictures in sequence to produce an animated display. This involves considerable computer resources to ensure that the speed of display is sufficient for the purpose.

To overcome the two-dimensionality of images, three-dimensional displays are being made available. These show the user a stereo image by interlacing two two-dimensional images, the eye positions of which have been displaced slightly to represent the human eye spacing. The interleaving can be carried out using a switchable polarized filter and special spectacles. In some systems the spectacles act as the filter and in others the spectacles are passive and the filter is attached to the display device.

One final feature of the display of results that is becoming available is the production of particle tracks. These show where the fluid particles travel within the flow domain. They are produced by

149

integrating the velocity data at a point to show where the particle will move to. Such displays are extremely useful in showing the qualitative features of a flow, such as vortices.

9.3 Checking a solution

When analysing the results of a simulation, certain pieces of information will be required. For example, we might need to know a prediction of the pressure difference between two points in the flow domain for some physical system. Then slight geometrical modifications might be made to the mesh and another CFD solution produced to find the comparable pressure difference for the modified geometry. Another requirement might be the investigation of the flow field structure at a series of places in the calculation domain. Whilst the user can run the solver, obtain converged numerical results and then find the required data, this is not a very satisfactory procedure. It is much better to add an intermediate step. This step is the determination of whether or not the solution produced by the CFD process is a reasonable one, i.e. it is of high quality and is likely to resemble the physical flow. Then, if the simulation is reasonable, the user can find the specific data that is required and have some confidence in the findings.

Some of the following features of a set of results can be used as checks on the quality of the results:

- The flow should look qualitatively correct. For example, it should flow in the directions that might be expected.
- Where boundary layers exist the results should show a velocity change that resembles that in a boundary layer. Near a stationary wall the velocity vectors should show that the velocity changes with the distance away from the wall. The velocity should be seen to fall from some value at a point well away from the wall, *the free stream value*, to zero at the wall. This should take place over several cells, perhaps five or more. If there are fewer cells than this inside the computed boundary layer, then the mesh is too coarse and should be refined near the wall.
- The mass of fluid entering the domain should equal the mass leaving the domain. This is often calculated by the program itself and reported in an ASCII file to the user.

150

- At points where the pressure is specified, the velocity field should be smooth. At these points the continuity equation is not satisfied and so fluid can leave or enter the domain in a non-physical way. If this can be seen to be happening it is clear that the fluid mass is not being conserved in overall terms.

If these simple checks show that there might be problems with the quality of the results, then users should consider checking their input data and changing their models, if necessary, before re-running the solver program.

9.4 Refining a computer model

If it looks likely that a model must be refined, a user must consider the advantages of producing a better prediction against the cost constraint of repeating the whole simulation process. Quite often even crude models can give large amounts of new and useful information to a user. This might prove adequate for the purposes of some users but not for others. It all depends on the application under consideration.

The process of refining a model might include any of the following:

- increasing the density of mesh points in a given area so that the changes of the flow variables in that area can be more accurately captured, for example in a boundary layer;
- improving the physics of the model, such as would happen if a more suitable turbulence model could be used.

In terms of effort, the first of these involves a large amount of work, as it involves rebuilding the mesh of the domain, either by repeating one of the mesh generation processes that are described in Chapter 6, or by using an adaptive meshing process. Once the mesh is built the fluid specification within the pre-processor and the setting of the boundary and initial conditions have to be carried out again, transforming the data generated as part of the original specification process onto the new mesh. Then, finally, the numerical control procedures have to be repeated before the solver can be run.

A systematic way of increasing the mesh density for a mesh with a regular structure is to double the number of cells in each of the local mesh directions. Similar refinement schemes can also be carried out with unstructured meshes by, for example, placing a new node at the centroid of each cell and then remeshing. With the new mesh a solution is calculated and the results obtained. When the results do not vary in global terms from one mesh refinement to the next, then the results are said to be 'mesh independent'. Whilst we would always like our results to be independent of the mesh size, for many industrial problems this is not always possible as the constraints in terms of cost or time or computer capacity are too great.

10 Some case studies

10.1 The examples

In Chapters 5 through 9 we have discussed the various stages of the CFD analysis process. Each of these chapters acts as a basic guide for an individual stage in the process. The time has now come to demonstrate how the whole process is used to produce a CFD simulation. To do this we will look at three examples that show the CFD analysis process being used. By going through these examples in considerable detail, it is hoped that the analysis process can be brought to life and some of the realities of carrying out the analysis process can be conveyed to the reader. The three examples that we will look at are the following:

- *A simple laminar flow.* To illustrate the basic procedures, we will look at predicting the two-dimensional laminar flow between two plates. Simple examples such as this are often used to test a new CFD program and to give the user some confidence that the program produces accurate results compared to known analytical solutions. Also, they can be used in the training of CFD users as they require very little computational effort to produce results.
- *The flow of air over a vehicle.* In this example, we simplify the three-dimensional problem of calculating the flow over a car by considering the flow to be in a two-dimensional plane corresponding to the vertical plane of symmetry. The flow is turbulent, however, and so we have to think about how to model the effects of the turbulence on the flow. To discretize the flow domain we use a mesh which has a regular topology, i.e. we use a structured mesh, but the mesh is distorted to fit

around the surface of the vehicle. Having looked at the two-dimensional problem, some of the results from three-dimensional simulations will be discussed, together with their implications for the use of CFD.

- *The flow of water around a combustion chamber.* This example considers a three-dimensional flow in a complex geometry, such as that found inside a water-cooled piston engine. Again, the problem is simplified in that the flow around a single combustion chamber is modelled. In this problem the turbulent flow through an inlet, a cooling chamber and an outlet is modelled as a fully three-dimensional flow using a mesh that is essentially unstructured.

These cases are described so that the reader can see exactly how the CFD solutions were produced by using commercial software and by following the analysis process that has been described in the preceding chapters. By studying these examples the reader should become more familiar with the tasks that need to be performed during the analysis; that is, the tasks of flow specification, mesh building, setting the fluid flow parameters, controlling the numerical solution, running the solver and analysing the results.

10.2 The software packages

All the cases have been run using commercial CFD software, although the meshes for the two turbulent flows have been built using simple, locally generated computer programs. For each case, the operating system commands that have been used to run the programs have not been given as these are often specific to a particular type of hardware; but the commands that have been used to set up the simulations within the software packages have been given. This has been done to give the reader a feel for the types of command that need to be issued, not to give a tutorial in the use of the software. In fact, the syntax of some of the commands will probably change before this book is published, and so the reader should be very wary of using the commands listed here. The reference guide or user manual of the particular CFD software package should always be consulted when creating the computer model for a simulation.

Two CFD software packages have been used to generate the flow simulations discussed here. The packages used are as follows:

- *PHOENICS*. This has been available since 1981 and is written by CHAM Ltd of Wimbledon. It is a CFD software package which uses the finite volume method to solve the governing equations on a staggered grid which has a regular topology. As it was one of the first packages to become available, it can be used to simulate a very wide variety of physical problems.
- *STAR-CD*. This is written by Computational Dynamics Ltd of London and is a CFD software package which uses the finite volume method to solve the governing equations on a non-staggered grid which can have an irregular topology. This capability to deal with an unstructured mesh is achieved by using the Rhie and Chow algorithm mentioned in Chapter 3.

These CFD packages have been used with the permission of the authors and it is not the intention of this book to draw comparisons between the two packages. Each of these packages has unique features and they are both used here solely to give a feel for how different packages can be used to produce flow simulations. In fact, if a user follows the simulation process that has been discussed, the CFD package might be thought of as being reduced to the role of a translator, translating the flow specification into a form that is understood by the solver program, and then translating the numerical results into a form understood by the user.

It must be recognized that the needs of users vary as people have to solve many different types of flow problem. This means that each user, or commercial organization, must decide what it is that they require from the use of a CFD software package. In every case the requirements that are decided upon will be different, but the process of making decisions can be standardized. This problem is addressed in Chapter 12, where the issues that determine the specification of a CFD package are discussed.

10.3 Laminar flow between parallel plates

10.3.1 *Producing the flow specification*

Figure 10.1 shows the flow situation for this simple example. We can see that two thin parallel plates, of length L m and distance h m apart, are placed horizontally in a flow which, well upstream of the

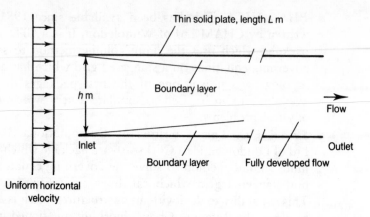

Figure 10.1. Flow between parallel plates

plates, has a constant velocity in the horizontal direction. If the flow velocity is sufficiently small or the kinematic viscosity sufficiently large, the Reynolds number will be low. If this is the case, then the flow should be laminar.

Given that this test case is being run as a simple training exercise, the first task in the production of a simulation is to consider what will happen to the fluid as it passes between the plates. First let us assume that the plates are so thin that the flow ahead of the plates is not affected by them. This means that we need only be interested in the flow between the plates, and that the flow above the top plate and below the bottom plate need not be considered. From this, the flow domain can be taken to be a simple rectangle. At the left hand side the flow has a uniform velocity in the horizontal direction moving from left to right, and so this boundary is an inlet. The plates are stationary solid walls and so the velocity there must be zero. Hence, there is a retardation of the flow at the plates due to viscous shear which is generated by friction, and two boundary layers are formed on the plates, as shown in Figure 10.1. These boundary layers become thicker along the plates from left to right until they merge. At the end of the plates, the fluid leaves the domain and so the right hand side of the rectangle may be taken to be an outlet.

This consideration of what happens enables us to see that the flow is symmetric about the horizontal plane half-way between the two plates, and so the flow domain can be halved for the purposes of our calculations. Figure 10.2 shows the rectangular domain and gives the four boundary types that will be used. These are a

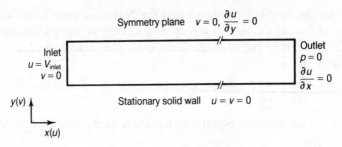

Figure 10.2. Domain with boundary conditions

stationary solid wall on the lower side where the velocity is zero, a symmetry plane on the upper side where the vertical velocity component is zero and the normal derivative of the horizontal velocity component is also zero, an inlet with a uniform horizontal velocity imposed at the left hand end, and an outlet where the pressure will be taken to be uniform at the right hand end.

We must also decide upon the values of the density and viscosity parameters. For simplicity, these will be taken to be unity in each case (i.e. $\rho = 1$ kg/m^3 and $\mu = 1$ kg/m s). Hence the Reynolds number Re is given by

$$Re = \frac{\rho V_{inlet} h}{\mu} = V_{inlet} h \qquad (10.1)$$

Finally, as the flow is a simple shear flow and none of the boundary conditions change with time, it is reasonable to assume that the flow itself will not vary with time and so will be steady. This completes the flow specification.

10.3.2 *Some analysis*

We have already said that this flow situation can be used as a test case to check the accuracy of a CFD code. This comes about because, some distance after the two boundary layers merge, the flow becomes one-dimensional. When this occurs the flow is said to be a fully developed flow, which means that the horizontal component of velocity does not change in the x-direction and that the vertical component of velocity is zero. If this flow is simulated using a mesh which is very long in the x-direction, then the CFD solver should produce results that are one-dimensional and the results should be of the form that will now be derived.

157

When the flow is fully developed the Navier–Stokes equations can be simplified. If the flow is steady and has the velocity characteristics given above, then the x-momentum equation (equation (2.8)) can be rewritten as

$$-\frac{\partial p}{\partial x}+\frac{\partial}{\partial y}\left(\mu\,\frac{\partial u}{\partial y}\right)=0 \tag{10.2}$$

and the y-momentum equation (equation (2.9)) can be rewritten as

$$\frac{\partial p}{\partial y}=0 \tag{10.3}$$

Equation (10.3) shows that the pressure is a function of x only, and so when equation (10.2) is integrated with respect to y the pressure derivative can be taken to be a constant. This gives

$$\mu\,\frac{\partial u}{\partial y}=\frac{\partial p}{\partial x}\,y+A \tag{10.4}$$

where A is a constant or a function of x only. Further integration with respect to y gives

$$\mu u=\frac{\partial p}{\partial x}\,\frac{y^2}{2}+Ay+B \tag{10.5}$$

where B is also a constant or a function of x. The values of A and B can be determined by applying the boundary conditions for the velocity at the two plates. We know that the horizontal velocity component u is zero at the plates, i.e. $u=0$ at $y=0$ and $y=h$, where h is the distance between the plates, and so equation (10.5) becomes

$$\mu u=\frac{\partial p}{\partial x}\,\frac{y^2}{2}-\frac{\partial p}{\partial x}\,\frac{hy}{2}=\frac{y}{2}\,(y-h)\,\frac{\partial p}{\partial x} \tag{10.6}$$

which describes a parabolic velocity profile.

Finally, we can calculate the mass flow in and out of the system. For an inlet velocity of 1 m/s and a density of 1 kg/m³, the mass flow per unit area is simply h and this must be the mass flow at the outlet too. Integrating the velocity expression in equation (10.6) to obtain the mass flow at the outlet:

$$h=\int_{y=0}^{h}u\;\mathrm{d}y=\frac{1}{\mu}\left\{+\frac{\partial p}{\partial x}\,\frac{h^3}{6}-\frac{\partial p}{\partial x}\,\frac{h^3}{4}\right\} \tag{10.7}$$

which can be rearranged to give an expression for the pressure gradient

$$\frac{\partial p}{\partial x}=-12.0\,\frac{\mu}{h^2} \tag{10.8}$$

Equation (10.8) enables the pressure gradient for a fully developed flow to be found for a given mass flow rate, and this can then be used in equation (10.6) to give the fully developed velocity profile for the same flow. These quantities can then be compared with the values calculated by the CFD program.

10.3.3 *Building a mesh*

Having produced a specification of the flow problem, we can now use a CFD program to produce a numerical simulation of the problem. The next step in this analysis process is to decide upon a suitable mesh and this part of the process is explained in Chapter 6. The domain and its boundaries are shown in Figure 10.2 and the mesh must fit within the domain in such a way that the variations in the flow variables can be calculated as accurately as possible.

For the flow situation that we are considering, we know that there is a boundary layer on each of the plates due to the shearing of the fluid caused by friction. We also know that at some distance downstream of the inlet, perhaps a factor of ten times the distance between the plates, the flow becomes fully developed and is effectively one-dimensional. Whilst this takes quite a distance to occur, the velocity changes most rapidly near the inlet. In the vertical direction, between the plates, the velocity profile is parabolic at the outlet and so it varies throughout the vertical distance.

For the problem that we are going to simulate we will take the distance between the plates h to be 1.0 m and the length of the plates to be 20.0 m. Hence, the computational domain is 0.5 m high and 20.0 m long. For our first mesh we will place ten cells between the lower plate and the symmetry plane, and ten cells down the length of the plates. To ensure that the rapid changes in velocity at the inlet can be captured, we will bias the mesh so that more cells are placed near the inlet. Between the plate and the symmetry plane we will use equal cell spacing, as we do not know where the velocity will vary the most in the vertical direction.

Having considered the layout of the mesh, we must work out how to create the mesh data in a form suitable for the CFD program. The way that this is done will be specific to the CFD software being used. For this example we will use the package PHOENICS, and so, before we look at the creation of the mesh data, we must first consider the software tools that make up PHOENICS.

PHOENICS is a finite volume program that is comprised of three main components or programs. The first of these, named the SATELLITE program, is a pre-processor; the second, named EARTH, is the solver program, and the third, PHOTON, is a graphical post-processor. Initially, the SATELLITE program has to be given sufficient information for it to produce the data that EARTH needs. One means of doing this is to prepare an input file for SATELLITE, which splits the input data into twenty-four groups. It is in this file that the mesh data is defined. SATELLITE can also be run interactively, allowing the user to create data or modify existing data in any of the twenty-four groups. When SATELLITE is run, files are produced for EARTH to read and, from this input data, EARTH produces the CFD solution in the form of further files, which are usually binary files, and these can be accessed using PHOTON. EARTH also produces some ASCII files which can be read by the user.

The input file for SATELLITE is known as the Q1 file, and several lines can be used to define a simple mesh. First we must decide upon the exact location of the mesh points. For this problem we will change the labels of the coordinate directions from x and y to z and y, respectively, that is we will take the direction between the plates to be the y-direction and the direction along the plates to be the z-direction. This choice of coordinate directions is determined by the internal structure of the programs that make up PHOENICS. These programs calculate the flow variables in sheets of points in the local z-direction, taking one sheet of points at a time. Here, the local z-direction and the global z-direction are the same (see Chapter 6). By carrying out the calculations in this way the number of points being considered at any one time is reduced from 100 to 10 for this problem. Whilst such a reduction is not significant for this mesh, as the memory storage requirements for this problem are small, it can mean the difference, when the mesh is much larger, between having sufficient computer memory to produce a solution and not having enough memory.

Returning to the generation of the mesh for this example, we have already stated that we will use equal cell spacing between the plates and so the mesh points will be at the following values of y: 0.0, 0.05, 0.1, 0.15, 0.2, 0.25, 0.3, 0.35, 0.4, 0.45 and 0.5. In the z-direction, the choice of points is more difficult, and so we will try values which halve the available distance, working from the outlet, i.e. we will use z-values of 0.0, 0.03906, 0.07813, 0.15625, 0.3125, 0.625, 1.25, 2.5, 5.0, 10.0 and 20.0. Once we have decided upon the mesh coordinates, we must build the mesh using appropriate commands in the Q1 file.

Below, we have listed an extract from the Q1 file. Note that where a line is indented by several spaces SATELLITE takes the line to be a comment not a command. The commands that specify the mesh are:

```
TALK=t;RUN( 1, 1);VDU=TTY
   GROUP 1.  Run  title  and  other  preliminaries
TEXT(SIMPLE DEVELOPING FLOW IN BETWEEN PLATES)
   A Cartesian coordinate system is used to encapsu-
   late a rectangular duct
   GROUP 2. Transience; time-step specification
   GROUP 3. X-direction grid specification
NX=1
   GROUP 4. Y-direction grid specification
NY=10
YFRAC(1)=0.05
YFRAC(2)=0.1
YFRAC(3)=0.15
YFRAC(4)=0.2
YFRAC(5)=0.25
YFRAC(6)=0.3
YFRAC(7)=0.35
YFRAC(8)=0.4
YFRAC(9)=0.45
YFRAC(10)=0.5
   GROUP 5. Z-direction grid specification
NZ=10
ZFRAC(1)=0.03906
ZFRAC(2)=0.07813
ZFRAC(3)=0.15625
ZFRAC(4)=0.3125
ZFRAC(5)=0.625
ZFRAC(6)=1.25
ZFRAC(7)=2.5
ZFRAC(8)=5.0
ZFRAC(9)=10.0
ZFRAC(10)=20.0
   GROUP 6. Body-fitted coordinates or grid distortion
```

Looking at the beginning of this extract from the Q1 file, there is a single line which tells the SATELLITE to read the Q1 file and then allow the user to modify the data interactively. Then the data is given group by group as follows:

Some case studies

- *Group 1 – Preliminaries*. A title for the simulation is given. This is printed on any ASCII files that are written by EARTH and on any pictures generated by PHOTON.
- *Group 2 – Time dependence*. Here, the transient nature of the problem can be specified, but PHOENICS assumes that problems are steady state unless told otherwise, and so there are no entries in this case.
- *Groups 3 to 5 – Mesh specification*. In these groups our simple mesh can be defined. For this problem, the x-direction is across the flow and so is not really needed for the simulation. However, PHOENICS is a program that must have a three-dimensional mesh and so there has to be a single cell in the x-direction. This is defined in Group 3 and the cell will have the default width of 1 metre. The mesh in the y-direction is specified in Group 4 by setting the number of cells (NY) to ten and by giving the coordinates of the mesh points. Similarly, in Group 5, the mesh in the z-direction is defined. This is all the information that is required to define the mesh.
- *Group 6 – Body-fitted coordinates*. As the mesh is very simple and not body-fitted, no entries are required here.

10.3.4 *Setting the fluid flow parameters*

Having defined the mesh, we can proceed with the next stage of the process and define the fluid flow problem. It is this part of the analysis process that tells the CFD software what the fluid properties are, together with the boundary conditions and the initial conditions. This section of the process is explained in Chapter 7 and involves translating the flow specification into terms understood by the CFD solver. Again, an extract from the Q1 file follows:

```
  GROUP 7. Variables stored, solved & named
SOLVE(P1,V1,W1)
  GROUP 9. Properties of the medium (or media)
ENUL=1.0
RHO1=1.0
  GROUP 10. Inter-phase-transfer processes and
           properties
  GROUP 11. Initialization of variable or porosity
           fields
FIINIT(W1)=1.0
```

162

```
GROUP 12. Convection and diffusion adjustments
GROUP 13. Boundary conditions and special sources
** Wall
PATCH(DUCTWALL,SWALL,1,1,1,1,1,NZ,1,1)
COVAL(DUCTWALL,W1,1.0,0.0)
** Inlet
PATCH(INLET,LOW,1,1,1,NY,1,1,1,1)
COVAL(INLET,P1,FIXFLU,1.0); COVAL(INLET,W1,ONLYMS,
1.0)
** Outlet
PATCH(OUTLET,HIGH,1,1,1,NY,1,NZ,1,1); COVAL(OUTLET,
P1,FIXVAL,0.0)
GROUP 14. Downstream pressure for PARAB=.TRUE.
```

It is these entries that determine the flow problem, and the commands that are entered group by group are as follows:

- *Group 7 – Solution variables*. We need to determine the variables that must be calculated. As this is a two-dimensional laminar flow problem, the equations to be solved are the momentum equations in the y- and z-directions, together with the continuity equation. The variables that we need to find to complete the solution are, therefore, the velocity components v and w, and the fluid pressure p. PHOENICS can solve problems that involve flows comprising several fluid components or phases, as discussed in Chapter 11, and so the entry here tells the software which variables to calculate by listing their names P1, V1 and W1. These variable names specify that the variables are those of the first fluid phase, which in this case is the only phase.

- *Group 9 – Properties*. We have already decided that the fluid viscosity and density should both be unity and these values are set here.

- *Group 10 – Multi-phase flows*. As this is a single phase flow, no extra specification is needed here.

- *Group 11 – Initialization*. In this group the initial conditions can be defined. The specification of these conditions might seem to be contradictory as the flow is steady and so, in a mathematical sense, the numerical solution should not require initial conditions. Despite this, we specify a set of initial conditions for the variables and these are then used as a first guess by the non-linear solution procedure. For this example we have defined the velocity component w to be

163

unity within the domain when the time is zero, and we let the other variables take their default value of zero.

- *Group 12 – Unused.* No entries.
- *Group 13 – Boundary conditions.* Finally, in specifying the fluid flow problem, we must specify the boundary conditions of the problem. This involves specifying where in the mesh the boundaries are and then applying the correct boundary conditions at the relevant boundaries. For this problem, the boundaries are shown in Figure 10.2, where we can see an inlet, an outlet, a solid stationary wall and a symmetry plane. To identify the location of the boundaries, PHOENICS uses the notation described in Section 7.3.1 and shown in Figure 7.1. Hence, the inlet is a LOW boundary, the wall is a SOUTH boundary, the symmetry plane is a NORTH boundary and the outlet is a HIGH boundary. The PATCH commands define the inlet, outlet and wall areas, giving the limits of a patch in the local coordinate directions x, y, z and the time t respectively. Note that the symmetry plane is not defined as this is the default boundary type in PHOENICS. So-called COVAL statements can then be used to apply the appropriate boundary conditions to the given boundary patches. On the wall the velocity component w is set to zero, at the inlet the mass flow and inlet velocity are specified and at the outlet the pressure is set to zero.

10.3.5 *Running the solution*

At this stage, all of the fluid mechanics parameters of this example have been defined, and so we can now set the parameters that control the numerical solution. The remainder of the Q1 file is listed below, and it can be seen that some of the groups are empty as no input is required and the default values will be used. The groups that do have entries are concerned with the control of the solver itself.

```
  GROUP 8. Terms (in differential equations) &
  devices
DIFCUT=0.5
  GROUP 15. Termination of sweeps
LSWEEP=100
RESREF(P1)=1.E-6;RESREF(V1)=1.E-6
RESREF(W1)=1.E-6
  GROUP 16. Termination of iterations
```

```
  GROUP 17. Under-relaxation devices
  GROUP 18. Limits on variables or increments to
  them
  GROUP 19. Data communicated by satellite to GROUND
  GROUP 20. Preliminary print-out
ECHO=F
  GROUP 21. Print-out of variables
  GROUP 22. Spot-value print-out
IXMON=1
IYMON=2
IZMON=2
  GROUP 23. Field print-out and plot control
IPROF=3
ITABL=3;NPLT=1
  GROUP 24. Dumps for restarts
STOP
```

The control commands for this problem are found in the following
groups:

- *Group 8 – Terms in the differential equations.* Here, only
  one parameter, DIFCUT, is specified. This determines the
  way in which the convection terms in the momentum equations are handled. For this simple laminar flow problem
  which is calculated on a rectangular mesh, the discretization
  of the convection terms should create very few problems. If
  the value of DIFCUT is set to 0.5, this tells EARTH to use
  a hybrid upwinding scheme, where the local cell Peclet
  number is calculated: if it is two or less, central differences
  are used; and if it is greater than two, upwinded differences
  are applied. Further details are discussed in Section 3.5.3.
- *Group 15 – Termination of sweeps.* In this group of data,
  the number of overall iterations, or sweeps as they are
  known to PHOENICS, is set to 100. Normally, far fewer
  iterations would be run to start the calculation and check its
  initial convergence performance, but for this problem the
  calculation is very robust and converges easily. The reference
  residual values or RESREF parameters are set such that the
  calculation will stop automatically if the value of the residual
  errors from the equations falls below the values specified.
- *Groups 20, 22 and 23 – Print out.* In these groups, the data
  that is written to an ASCII file is controlled. The ECHO

command suppresses the printing of the data read by EARTH and the rest of these commands ensure that the residuals are printed to the file at each iteration, together with the values of the velocity components and pressure at one cell in the mesh. This cell has been chosen to be near the inlet so that the variation in the variable values can be monitored as the iterations progress. It is known as a *monitor location*. As well as numerical values, simple graphs of the spot values and residuals are printed.

To run this model the following stages are carried out. First, the SATELLITE program reads the Q1 file and then allows the user to check the settings of the data in each of the groups interactively. Once the user is satisfied that everything is in order, the SATELLITE program writes the datafiles that EARTH requires. Then EARTH is run and produces a set of output files. Some of these can be read into PHOTON for graphical analysis or the ASCII files can be read by the user using a browse facility or text editor.

10.3.6 *Analysing the results*

In Section 10.3.2, we derived some analytical results for the flow under consideration. These are valid near the outlet of the flow where the velocity field is fully developed and one-dimensional. Consequently, near the outlet, we can determine the exact values of the pressure gradient and the velocity profile that should be calculated by the CFD solver program. From equation (10.8), if we substitute for the viscosity and domain height, the pressure gradient can be found to be

$$\frac{\partial p}{\partial x} = -12.0 \qquad\qquad (10.9)$$

and the velocity profile can be found from equation (10.6) as

$$u = 6.0(y - y^2) \qquad\qquad (10.10)$$

By comparing the output from PHOENICS with the expressions above, we can obtain some measure of accuracy for our numerical solution.

For this test case the mesh contains only 100 cells and so it is a manageable task to read the output files from PHOENICS in full. An edited version of the output written by EARTH is listed in the Appendix. This output consists of the mesh point data, the final

166

values of the variables calculated within each cell, the variation of the values of the variables at the monitor location with sweep number and the variation of the residuals with sweep number.

From the data listed in the Appendix it can be seen that, after the first 100 sweeps, the following information is available:

- From the pressure field data and the locations of the mesh points, the pressure gradient near the outlet is -9.564. This has been calculated manually, knowing that the pressure is stored at a cell centroid.
- From the velocity field data, the W1 velocity component is varying all the way down the mesh in the z-direction and so the flow is not fully developed at the outlet.
- At the monitor location, the value of the pressure P1 is rising steadily, the velocity component V1 has risen to a peak and is now falling and the velocity component W1 has fallen to a minimum and is now rising.
- The residual errors are all falling, but those for V1 and W1 rose initially before falling.

The data calculated for the flow field suggests that the numerical solution is not that which is expected and the data from the monitor location shows that the solution is not converged. However, as the residuals are falling, the solution is progressing satisfactorily and further sweeps need to be run to see whether a solution will be produced which is numerically converged and also closer to our expectations. To do this, a restart calculation has to be performed. This is done by running SATELLITE again, and telling EARTH to use the values of the variables that have previously been calculated. The necessary command is RESTRT(V1,W1,P1) and this is entered in Group 11. As 100 sweeps have not produced a converged solution, the value of LSWEEP is also increased to 400. Once SATELLITE has written a new set of datafiles, EARTH can be run to produce a further solution.

Again, the Appendix has an edited listing of the PHOENICS output after what now amounts to 500 sweeps. The solution now yields the following information:

- From the pressure field data, the pressure gradient is now -12.48.
- From the velocity field data, the W1 velocity component is approximately constant with distance down the plates in the

167

columns of cells numbered 6, 7 and 8 in the z-direction. This shows that the flow is becoming fully developed near the outlet but the process is not quite complete as yet.
- At the monitor location, the value of the pressure P1 is rising but at a reducing rate and so can be seen to be converging. The velocity component V1 is again rising but converging, and velocity component W1 has fallen to a minimum and is now rising, if only slowly.
- The residual errors are all falling.

This solution is clearly a much better one, the solution is converging and the actual values are looking like those we would expect, even if they are not quite right. To improve the situation still further, or at least to try to, we can run the solution for another 400 sweeps, restarting from the latest solution. Again, the Appendix contains the results after this additional calculation and from these the following can be seen:

- From the pressure field data, the pressure gradient has changed slightly to − 12.55.
- From the velocity field data, the W1 velocity component is approximately constant with distance from the plates in the columns of cells numbered 6, 7, 8 and 9 in the z-direction and so the flow is effectively fully developed near the outlet.
- At the monitor location, the value of the pressure P1 is rising but at a reducing rate and so can be seen to be still converging. The velocity component V1 is falling but converging, and velocity component W1 is rising and converging. In fact, the changes are magnified in the plot of the values, as, if we look at the last few sweeps, only the fourth significant figure is changing.
- The residual errors are all falling.

Now the solution is effectively converged and the accuracy of the simulation can be calculated. Table 10.1 shows the outlet velocity at the end of each of the three solution runs together with the analytical solution from equation (10.10). From this it can be seen that there is only a small error, which is worse near the wall.

PHOTON can now be used to produce pictures of the flow from the results. In Figure 10.3 the velocity vectors can be seen, and this picture has been created by using the following PHOTON

Table 10.1 *Simple flow - velocity comparison*

y	Analytical	Sweep 100	Sweep 500	Sweep 900
0.025	0.146	0.464	0.127	0.109
0.075	0.416	0.612	0.393	0.388
0.125	0.656	0.756	0.639	0.639
0.175	0.866	0.890	0.857	0.858
0.225	1.046	1.011	1.043	1.047
0.275	1.196	1.116	1.199	1.204
0.325	1.316	1.201	1.323	1.329
0.375	1.406	1.266	1.417	1.423
0.425	1.466	1.309	1.479	1.486
0.475	1.496	1.376	1.510	1.517

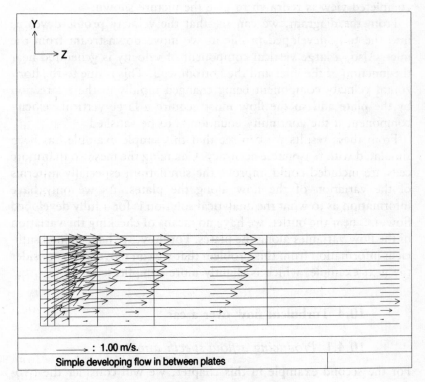

Figure 10.3. Parallel plates – velocity vectors

169

commands:

```
grid x 1
magnify grid 10
vector x 1
set reference vector
1.0
redraw
```

When PHOTON is used, the default view has the *y*-direction as the up-direction and the *z*-direction goes from left to right. This is exactly the orientation that we require and so no commands are required to specify it. The first command that is given draws one sheet of cells onto the screen and these are then magnified by a factor of ten with the cursor being used to put the centre of the screen near the inlet. Then the vectors are drawn and the reference vector set so that the length of a typical vector is as shown. This prevents the vectors being extremely long and filling the screen. Finally, the completed view is redrawn to give the picture shown.

From the diagram, we can see that the velocity profile develops into the fully developed profile as we move downstream from the inlet. Also, a large vertical component of velocity is generated near the junction of the inlet and the bottom wall. This is due to the horizontal velocity component being changed rapidly in the *z*-direction by the plate and so the flow must acquire a large vertical velocity component if the continuity equation is to be satisfied.

From these results we can see that this simple example has been simulated with reasonable accuracy. Changing the mesh so that more cells are included could improve the simulation, especially in terms of the variation of the flow along the plates. As we only have information as to what the analytical solution is for a fully developed flow, i.e. near the outlet, we have no means of checking the variation of the flow variables along the plates. For now, we have gained sufficient information from this problem that we can move on to consider the next example, which is slightly more complicated.

10.4 Turbulent flow over a car

10.4.1 *Producing a flow specification*

For the second example in this chapter, we will consider the two-dimensional situation that we have discussed already in Chapter 5.

This example involves the simulation of the turbulent flow over the longitudinal section of a car. In Chapter 5, we considered this flow in some detail, producing the flow specification of the problem, and so we can proceed immediately to build the computer model of this flow. The software that we will use to produce this simulation is PHOENICS once again, and the structure of this package has already been discussed. We will use this software in the same way as we did for the first example, but we must now consider modelling a turbulent flow as well as fitting the mesh to the surface of the vehicle. As a matter of personal preference, the mesh will be produced outside of the PHOENICS program, using locally written software, but the mesh generation tools of PHOENICS itself could also be used.

10.4.2 *Creating a mesh*

From our consideration of the flow during the flow specification phase, we know that the flow variables will vary greatly in boundary layers near the vehicle surface and on the other solid walls that make up the wind tunnel. This means that we need to be able to produce a mesh which has many cells near the vehicle surface. At the same time, however, PHOENICS must be provided with a mesh that has a regular topology, i.e. a structured mesh. One way of creating a structured mesh, for this example, is to build the mesh in nine parts or blocks, as was shown in Figure 6.5. As the full mesh must have a regular structure, so the mesh in each block must also have a regular structure and the distribution of the cells within each block must be such that the cell faces match those of other blocks.

Such a mesh can be defined in the following two stages:

- First, a set of points on the vehicle surface must be calculated. These points are created such that they define the front, top, bottom and rear of the fifth block of cells shown in Figure 6.5.
- Second, points are created in each of the nine blocks using the points on the vehicle surface and the known geometry of the wind tunnel.

The shape under consideration here is a two-dimensional section of a full size model vehicle that has been used extensively to investigate and compare the wind tunnels used by vehicle manufacturers. The model vehicle is placed in the wind tunnel being tested and various forces such as the aerodynamic drag on the vehicle

171

found, together with the associated moments of these forces. As well as measuring these forces and moments, engineers can use the model to measure the surface pressure on the vehicle as there is a series of holes along the centreline section and around the waist of the model. Hence, there is an extensive database of flow data for this model which can be used to validate CFD codes. These uses of both the model itself and the data are well documented (see Carr, private communication; Shaw, 1988; Shaw and Simcox, 1988; Hawkins *et al.*, 1990), as is the shape of the vehicle.

From the drawings of the vehicle, the coordinates of the centreline section of the vehicle can be computed, as the three-dimensional surface is comprised of planes, together with cylindrical and spherical sections. A simple program has been written to produce the surface coordinates for the centreline section of the vehicle. This program is given the number of cells that there will be on the vehicle surface in the flow direction, which we will take to be the global z-direction, and the number of cells in the vertical direction, the global y-direction. The coordinates of the points on the top of the computational block, i.e. the bonnet, the windscreen, the roof, the rearscreen and the boot, and the bottom of the block, the vehicle undersurface, are then calculated by the program at the values of z that it is given. On the front and rear surfaces of the block, the coordinates of the points are found for a set of y values calculated by the program using a cosine distribution.

Once the surface coordinates are known, the points within each of the blocks can be built up. This is done by a second program which reads the surface coordinates, together with the position of the wind tunnel inlet, floor, roof and outlet, and the number of cells in each block in the two directions z and y. Points are placed along horizontal and vertical lines, as appropriate, within the eight blocks outside the vehicle surface. Figure 10.4 shows a simplified mesh of the domain and from this the point creation algorithm can be deduced. Above the vehicle, in block 6, and below it, in block 4, vertical lines are created from the points on the vehicle surface to the tunnel roof or floor. Similarly, ahead of the vehicle, in block 2, and behind it, in block 8, horizontal lines are created from the points on the vehicle surface to the tunnel inlet and outlet. Then the coordinates of the points that form the cell corners are found by splitting each line into sections using a geometrical progression to bias the positioning of the points.

For simple flows, where the variation of the flow variables is not

172

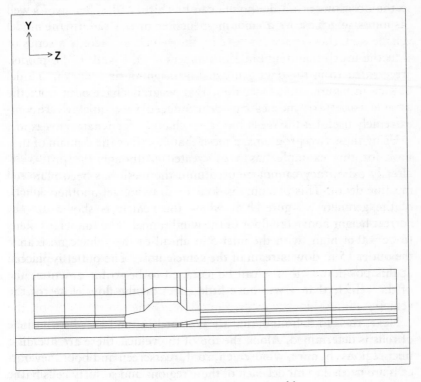

Figure 10.4. Simplified mesh for the car problem

too great, the position of each point could be found by splitting the line into equal intervals, given the number of cells that need to be placed along the length of the line. As we need to be able to describe boundary layers along the vehicle surface, geometrical progression biasing is used to create the points along the line such that there are more points near the vehicle surface. For a geometric progression, the sum of n terms is given by

$$S_n = a + ar + ar^2 + \cdots + ar^{n-1} = \frac{a(1 - r^n)}{(1 - r)} \tag{10.11}$$

where S_n can be taken to be the length of the line, a is the length of the first interval and r is the ratio of neighbouring element lengths.

The mesh generation program is given the ratio of the length of the element near the tunnel boundary to the length of the element at the vehicle surface, and then computes the value of the ratio r. From equation (10.11) the length of the first element a can be found and

173

so the positions of all the points can be calculated. This gives a set of points which show a smooth reduction in cell size towards the vehicle surface.

In the four remaining blocks, numbers 1, 3, 7 and 9, the points are created from the data generated in neighbouring blocks, as can be seen in Figure 10.4. Once these two programs have been written, a wide variety of meshes can be produced very quickly. This is extremely useful if the mesh has to be changed for whatever reason.

Using these two programs, a mesh that describes the domain of the flow for this example has been created. Although the programs already exist, they cannot be used until the mesh has been planned in some detail. This planning is done by drawing yet another sketch of the geometry, Figure 10.5, where the vehicle is shown at the correct height above the floor of the wind tunnel. The tunnel is taken to be 3.0 m high, with the inlet 5 m ahead of the vehicle nose and the outlet 15 m downstream of the vehicle nose. The outlet is placed at this position so that it can be assumed to be so far downstream of the vehicle that it will have little effect on the flow close to the vehicle.

Next, the block boundaries are sketched in, and the distribution of cells is determined. Along the top of the vehicle there are five distinct regions: bonnet, windscreen, roof, rearscreen and boot. Several cells are needed to model each of these regions and so forty cells have been placed along the whole vehicle length. The positions of the cells have been chosen so that there are more cells near the boundaries of the regions on the top of the vehicle. Ahead of, and behind, the vehicle, the flow changes rapidly near the car and it changes very

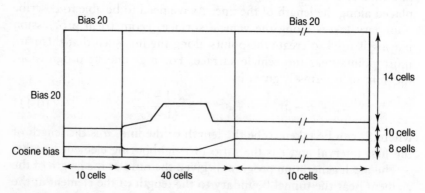

Figure 10.5. Sketch of a mesh layout for the car problem

little near the inlet and outlet. Some ten cells have been placed in the horizontal direction in these areas and the cells are biased so that the cells nearest the vehicle are twenty times shorter than those near the inlet or outlet. This value of twenty is a first guess for the biasing required to produce a reasonable simulation.

In the vertical direction, eight cells have been placed between the vehicle and the tunnel floor, with the cell sizes being determined by a cosine distribution. This makes the cell size smaller near the vehicle surface and the floor to allow the gradients in the boundary layers to be captured. Through the height of the vehicle ten cells have been placed and fourteen cells have been placed between the tunnel roof and the vehicle roof. The positions of the cells above the vehicle have been chosen such that the cell size at the vehicle is twenty times smaller than the cell size at the tunnel roof. Of course, there will be a boundary layer on the tunnel roof and this distribution of cells will not be able to describe the flow variation there. For this simulation, we have assumed that the roof is so far from the vehicle that it will not affect the flow around the vehicle, and so a symmetry boundary condition will be used. This will ensure that the tunnel roof will constrain the flow during the simulation by acting as a frictionless solid boundary.

This completes the specification of the mesh, and the mesh generation programs produce a datafile which contains the number of cells in the local coordinate directions, together with the coordinates of the corner points of the cells. This file is read by the SATELLITE program of PHOENICS using commands discussed in the next section.

10.4.3 *Preparing the data before solution*

Once a mesh has been created, the input file to SATELLITE, the Q1 file, has to be created. It is this file that is read by the SATELLITE program before it prepares the data for the EARTH program. This section discusses both the commands required to specify the flow problem and the commands required to control the numerical solution process. The commands are arranged in groups to assist the user in setting up the computer model, enabling small sets of data to be handled at any one time. A full listing of the Q1 file is given below followed by a description of the commands in each of the twenty-four groups, listed group by group.

```
TALK=T;RUN( 1, 1);VDU=TTY
   GROUP 1. Run title and other preliminaries
TEXT(TWO-DIMENSIONAL MOTOR VEHICLE)
REAL(W1IN,KEINIT,EPINIT)
   GROUP 2. Transience; time-step specification
STEADY=T
   GROUP 3. X-direction grid specification
NX=1
   GROUP 4. Y-direction grid specification
NY=32
   GROUP 5. Z-direction grid specification
NZ=60
   GROUP 6. Body-fitted coordinates or grid distortion
BFC=T
NONORT=T
READCO(GRID)
RSTGEO=F
SAVGEO=T
   GROUP 7. Variables stored, solved & named
SOLVE(V1,W1)
SOLUTN(P1,Y,Y,Y,N,N,N)
STORE(UCRT,VCRT,WCRT)
   GROUP 8. Terms (in differential equations) &
           devices
DIFCUT=0.0
ADDDIF=T
   GROUP 9. Properties of the medium (or media)
ENUL=1.46E-05
RHO1=1.225
TURMOD(KEMODL)
   GROUP 10. Inter-phase-transfer processes and
           properties
   GROUP 11. Initialization of variable or porosity
           fields
W1IN=28.0
KEINIT=0.005*W1IN*W1IN
EPINIT=944.0
FIINIT(KE)=KEINIT
FIINIT(EP)=EPINIT
FIINIT(P1)=0.0;FIINIT(W1)=W1IN;FIINIT(V1)=0.0
CONPOR(INTERIOR,0.0,CELL,1,1,9,18,11,50)
   GROUP 12. Convection and diffusion adjustments
   GROUP 13. Boundary conditions and special sources
PATCH(INLET,LOW,1,NX,1,NY,1,1,1,1)
COVAL(INLET,W1,ONLYMS,W1IN)
COVAL(INLET,P1,FIXFLU,W1IN*RHO1)
```

```
COVAL(INLET,KE,ONLYMS,KEINIT)
COVAL(INLET,EP,ONLYMS,EPINIT)
PATCH(FRONT,HWALL,1,1,9,18,10,10,1,1)
COVAL(FRONT,V1,GRND2,0.0)
COVAL(FRONT,KE,GRND2,GRND2)
COVAL(FRONT,EP,GRND2,GRND2)
PATCH(BOTTOM,NWALL,1,1,8,8,11,50,1,1)
COVAL(BOTTOM,W1,GRND2,0.0)
COVAL(BOTTOM,KE,GRND2,GRND2)
COVAL(BOTTOM,EP,GRND2,GRND2)
PATCH(REAR,LWALL,1,1,9,18,51,51,1,1)
COVAL(REAR,V1,GRND2,0.0)
COVAL(REAR,KE,GRND2,GRND2)
COVAL(REAR,EP,GRND2,GRND2)
PATCH(TOP,SWALL,1,1,19,19,11,50,1,1)
COVAL(TOP,W1,GRND2,0.0)
COVAL(TOP,KE,GRND2,GRND2)
COVAL(TOP,EP,GRND2,GRND2)
PATCH(FLOOR,SWALL,1,1,1,1,1,NZ,1,1)
COVAL(FLOOR,W1,GRND2,0.0)
COVAL(FLOOR,KE,GRND2,GRND2)
COVAL(FLOOR,EP,GRND2,GRND2)
PATCH(OUTLET,HIGH,1,1,1,NY,NZ,NZ,1,1)
COVAL(OUTLET,P1,FIXP,0.0)
   GROUP 14. Downstream pressure for PARAB=.TRUE.
   GROUP 15. Termination of sweeps
LSWEEP=10
   GROUP 16. Termination of iterations
LITER(P1)=20;LITER(V1)=1;LITER(W1)=1;LITER(KE)=1;
LITER(EP)=1
   GROUP 17. Under-relaxation devices
RELAX(P1,LINRLX,0.1)
RELAX(W1,FALSDT,0.006)
RELAX(V1,FALSDT,0.006)
RELAX(KE,FALSDT,0.0005)
RELAX(EP,FALSDT,0.0005)
KELIN=1
   GROUP 18. Limits on variables or increments to them
   GROUP 19. Data communicated by satellite to GROUND
   GROUP 20. Preliminary print-out
   GROUP 21. Print-out of variables
OUTPUT(W1,Y,N,N,Y,Y,Y)
OUTPUT(V1,Y,N,N,Y,Y,Y)
OUTPUT(P1,Y,N,N,Y,Y,Y)
OUTPUT(UCRT,Y,N,N,N,N,N)
OUTPUT(VCRT,Y,N,N,N,N,N)
```

```
OUTPUT(WCRT,Y,N,N,N,N,N)
   GROUP 22. Spot-value print-out
IXMON=1;IYMON=20;IZMON=39
   GROUP 23. Field print-out and plot control
NPLT=1;ITABL=3
   GROUP 24. Dumps for restarts
SAVE=T
   RESTRT(V1,W1,P1,KE,EP)
   SAVGEO=F
   RSTGEO=T
LSWEEP=150
STOP
```

The first line of the Q1 file tells SATELLITE to allow both interactive checking and modification of the data once the file has been
read. It also determines the computer terminal type that will be used.
Then the commands that specify the structure of the mesh, read the
previously prepared mesh data and set up the flow problem are given
in the following groups:

- *Group 1 – Preliminaries.* This contains a simple title and a
  list of user-defined variables that SATELLITE needs to know
  are real numbers. These variables will be used later in the Q1
  file as part of some simple calculations.
- *Group 2 – Time dependence.* Here, the flow is specified as
  being steady state, i.e. there is no variation with time. This
  is a simplification of the problem, made so that a solution
  can be found using a reasonable amount of computer time.
  In reality there is always some time variation of a turbulent
  flow, but we hope that for our computation the turbulence
  model will take this into account.
- *Groups 3 to 5 – Mesh specification.* This is where the
  program is told how many cells there are in each of the three
  local mesh directions.
- *Group 6 – Body-fitted coordinates.* The program is told that
  a mesh has been created that is body-fitted and that the
  coordinates of the grid points are stored on a file called
  GRID. As the mesh has been created using projections of
  points in the vertical and horizontal directions, it is clear that
  no attempt has been made to ensure that the mesh is orthogonal. EARTH needs to know this as extra numerical terms
  must be used in the numerical analogue of the governing
  equations when the grid is non-orthogonal. Once EARTH

178

has read the set of grid points, it can create a file which contains a great deal of geometrical information within it. To save time when performing a simulation, this file need only be created once and then stored. As the Q1 file listed refers to the first run of a solution, the last two commands in this group tell EARTH that the geometry file does not exist and that it should save this file at the end of this run.

- *Group 7 – Solution variables.* To solve this problem we need to find two velocity components and the pressure of the fluid. The velocity components, variables V1 and W1, are calculated by the program in directions defined locally in each cell, and these directions are determined by the positions of the corner points of a cell. As the post-processor PHOTON needs to have access to the velocity components defined in the Cartesian directions, we must calculate and store these additional components. These velocity components are known as UCRT, VCRT and WCRT. The command SOLUTN is used to activate the pressure variable P1 so that the default slab-by-slab solution method is changed to a whole-field solution method. This does not affect the values of the solution, but it does speed the solution process up.

- *Group 8 – Terms in the differential equations.* Here, the convection operator is requested to be formed using upwind differences regardless of the value of the cell Peclet number. This is done by setting the value of DIFCUT to zero. The ADDDIF command ensures that the pressure correction equation (see Section 3.5.2) includes the diffusion terms in the momentum equation and does not leave them out, as it would by default. This inclusion of terms increases the likelihood of the solution converging.

- *Group 9 – Properties.* As the vehicle is in air, the density and laminar kinematic viscosity are set to the values determined during the flow specification stage, Section 5.2. The TURMOD command switches on the two-equation k–ε turbulence model, telling EARTH to solve for both turbulent kinetic energy k and the rate of its dissipation ε, calculating the effective turbulent viscosity using the relationship given by equation (2.18) in Section 2.2.3.

- *Group 10 – Multi-phase flows.* No entries.

- *Group 11 – Initialization.* This group is used to define the value of the velocity, turbulent kinetic energy and dissipation

179

rate at the inlet. At the inlet boundary, the velocity component of the flow in the local z-direction, W1, the direction of
which coincides with the global z-direction at the inlet, is
28 m/s. This velocity is convected into the domain together
with the turbulent kinetic energy and its dissipation rate. To
calculate the values of the turbulence quantities at the inlet,
the value of turbulence intensity is assumed to be 6 per cent.
From the definitions of turbulence intensity I and turbulence
kinetic energy k (Schlichling, 1979; Bradshaw, 1971), the
value of the turbulence kinetic energy can be calculated as

$$k = \tfrac{3}{2} I^2 V_{inlet}^2 = \tfrac{3}{2}(0.06)^2 V_{inlet}^2 = 0.005 V_{inlet}^2 \qquad (10.12)$$

The ε-value is set so that it gives an effective turbulent kinematic viscosity which is 100 times the laminar kinematic viscosity, a typical value for air. Hence using equation (2.18)
with the coefficient c_μ set to its standard value of 0.09 we
obtain

$$\varepsilon = \frac{0.09 k^2}{\nu_T} = \frac{(0.09)(3.92^2)}{1.465 \times 10^{-3}} = 944 \text{ J/m}^3 \text{ s} \qquad (10.13)$$

These values will also be used when the boundary conditions
are prescribed in a later group, but they are calculated here
so that they can be used as the first guess to the variables
throughout the field. The FIINIT commands set the value of
the pressure and the velocity component V1 to zero, and of
the velocity component W1 and the k and ε variables to the
values at the inlet discussed above. These FIINIT commands
work by setting the values of the variables at every location
to the appropriate numerical value.

Finally, in this group, the cells inside the vehicle, which
should not take any part in the simulation as they exist only
to assist in the computational housekeeping, are labelled and
switched off using the CONPOR command.

- *Group 12 – Unused.* No entries.
- *Group 13 – Boundary conditions.* Here the boundary conditions are defined. We have already identified the boundaries as being physically located at the inlet, outlet, floor
and roof of the tunnel, and at the vehicle surface. As we have
decided to make the roof of the tunnel a symmetry plane and
not a viscous wall, we do not need to do anything to apply
this boundary condition as this is the default boundary condition. The vehicle surface can be described as the top of the

car together with the bottom, the front and the rear. These four sections of the surface form the boundaries of the fifth block used to create the mesh. The PATCH commands define the positions of the boundaries by listing the cell ranges and the face positions using the compass notation described in Section 7.3.1, and the COVAL statements apply the appropriate boundary conditions. These conditions are specified such that at the inlet the velocity component in the z-direction and the values of k and ε are specified, together with a mass flow boundary condition for the pressure correction equation (see Sections 3.5.2 and 3.5.4); at the outlet the pressure is set to zero; and on the other boundaries, wall functions are used to find the values of k, ε and the necessary velocity components.

Now that the information defining the fluid flow problem has been explained, the rationale behind the choice of the settings for the control parameters relating to the numerical solution must also be explained group by group:

- *Group 14 – Parabolic flow*. No entries.
- *Group 15 – Termination of sweeps*. The number of sweeps is set to ten. This enables the initial progress of the residual errors to be monitored to see if the solution process is moving in a satisfactory way towards convergence. If the residual values fall then the process is proceeding well; but if the residuals get bigger then the solution process may not converge.
- *Group 16 – Internal iteration control*. Within each sweep, the calculation of each of the variables involves the solution of a set of simultaneous linear equations. These solutions are found by an iterative procedure inside EARTH. It is important that the solution to the pressure correction equation is computed as accurately as is realistically possible, as this ensures that the mass of fluid is conserved throughout the flow domain, and so twenty internal iterations are performed when calculating the pressure. For the other four variables, accuracy at the end of a sweep is less important, and so only one iteration of the linear equation solver is performed per sweep.
- *Group 17 – Relaxation parameters*. To control the non-linear solution procedure, the variables that are calculated

181

must be relaxed in some way. For pressure this is done using standard linear relaxation, equation (8.3), and the relaxation factor is set to 0.1. The other variables are relaxed using a form of time step smoothing which is sometimes called pseudo-time relaxation. Although this problem is being calculated as a steady state problem, PHOENICS allows the addition of a false time-dependent term that smooths the solution procedure. Effectively, a time derivative of a variable, $\dot{\phi}$, which has the form

$$\dot{\phi} = \frac{\phi_{\text{new}} - \phi_{\text{old}}}{\Delta t} \tag{10.14}$$

is added to the left hand side of the discretized momentum equations. When the solution is converged this term will be zero, but in the initial stages it provides a smoothing of the solution. The value of the time step Δt is found by calculating a typical residence time for a particle in a cell. Here, an average velocity w in the z-direction is 28 m/s; there are sixty cells N_c in the z-direction covering a distance L of 20 m; and so an estimate of the residence time is given by

$$t_{\text{res}} = \frac{L}{N_c w} = \frac{20}{(60)(28)} = 0.0119 \text{ s} \tag{10.15}$$

For the velocity components a suitable, and conservative, value of Δt is about half this residence time, i.e. 0.006 s. The turbulence variables k and ε require more relaxation than this and so a value about one-tenth of relaxation for the velocity components is used, i.e. 0.0005 s. These values are only first guesses, chosen by what is a useful rule-of-thumb. In practice the required values depend on the shape of the mesh and the flow itself, and so some modification to these values may well be required.

Finally, in this group, the form of the $k-\varepsilon$ model is made appropriate for an external flow using the KELIN statement. This selects one form of the linearization of the terms in the $k-\varepsilon$ model, which the PHOENICS reference manual suggests should be suitable.

- *Groups 18, 19 and 20 – Special features and printout.* No entries.
- *Groups 21, 22 and 23 – Printout.* In these groups the output from EARTH is specified. In particular, graphs and lists of the residual errors are requested, together with graphs and

lists of the variables at a monitor location. This location is chosen to be near the upper surface of the vehicle where the flow varies rapidly in space. At this position, the change in the variables from sweep to sweep should provide a sensitive measure of the convergence of the numerical solution.

- *Group 24 – Restart data.* Here, the results are stored and the marker denoting the end of the Q1 file is written. The indented commands do not really belong in this group, but they are conveniently located at the end of the file as they are the commands that need to be activated if a restart solution is to be performed. They tell EARTH to read the last set of results and use them as the initial values for the continuation of the solution, to read the existing file that contains the geometry data and not to rewrite this file at the end of the solution.

Once the input data has been assembled and written in the Q1 file, the SATELLITE program can be run. This produces input data suitable for being read by the EARTH program, which can then be run in turn.

10.4.4 *Running the solver and analysing the results*

To check that the values of the relaxation parameters are suitable and that the solver produces results which appear to be converging, only a few sweeps have been run. As the residuals decreased from sweep to sweep during this trial run, the Q1 file was edited to instruct EARTH to run 150 sweeps. By running the SATELLITE, a new set of datafiles for EARTH were written, and EARTH itself run again. At the end of its run EARTH produced a report file, known as RESULT, in ASCII format. This contains various information, including the reports that are requested using the Q1 file. Amongst these results are the values of the variables within the monitor location cell at the end of each sweep. These are given in both numerical and graphical form. Similarly, the values of residuals for each of the equations are listed in the same way. Figure 10.6 shows the graph that EARTH has produced of the monitored values against sweep number and Figure 10.7 shows the graph of the residual error against sweep number.

In both these diagrams, the abscissa is the sweep number from 1 to 150, scaled to be in the range 0 to 1. Also, the spot values and

Some case studies

```
SPOT VALUES VS. SWEEP
     IXMON=      1   IYMON=     20   IZMON=      39

 1.00 V..E.+....+....+....+....+....+....+....+....+....+....+
      . EEK                                                  .
 0.90 +PEWE                                                  +
      . E  EW                                                .
 0.80 +  K  EK                                               +
      .EW   EKK                                              .
 0.70 WE    EEKK                                             +
      . W    EEK                                             .
 0.60 +K      EK          VVVVVV                             +
      EEP    EEK VVVVVVVV    VVVVVVVVVVVVVVVVVVVVVVVVVVVVVVV
 0.50 +K     EK VV                                           +
      K P    WEKV                                            .
 0.40 E P      EK                                            +
      .  P      VWEK                                         .
 0.30 +V P     VWEK                                          +
      W VPP    V  WEK                                        PP
 0.20 K V P   VV  EEKPPPPPPPPPPPPPPPPPPPPPPPPPPPPPPPPPPPPPPPP
      .VV PVVV   WEEKK                                       .
 0.10 +   VV    PPWWEEEK                                     +
      . VVPP  PP  WWWEEEE      WWWWWWWWWWWWWWWWWWKKKKKK
 0.00 +..V.+PPPP+....+...WEEEEEEEEEEEEEEEEEEEEEEEEEEEEEEEEEE
      0   .1   .2   .3   .4   .5   .6   .7   .8   .9  1.0
THE ABSCISSA IS     ISWP.  MIN= 1.00E+00 MAX= 1.50E+02
```

Figure 10.6. Monitor values against sweep (bias = 20)

```
RESIDUALS VS. SWEEP
 1.00 EW...+....+....+....+....+....+....+....+....+....+....+
      EEV                                                    .
 0.90 +EWV                                                   +
      P EWV                                                  .
 0.80 +PEEWVV                                                +
      VP EEW V                                               .
 0.70 + P  EEW VV                                            +
      .  P  EEWWVVV                                          .
 0.60 W   P EEWWWWVWV                                        +
      .   PP EE WWWW VVV                                     .
 0.50 +    PP EE  WWWWV                                      +
      .      P KEE   WWWV                                    .
 0.40 +     PP EEE    WWW                                    +
      .      PP  EEE   WWWWW                                 .
 0.30 +     PPP  EEE     VVWWWWW                             +
      .      PPPP EEEE    VVV  WWWW                          .
 0.20 +        PPPPPEEEEE    VVV   WWW                       +
      .         PPPP KEEEEEE VVVV  WWWW      W               .
 0.10 +            PPPPPPKKEEEEEEEEVV   WWWW     +
      .                 PPKKEEEEEEEEEWWWW                    .
 0.00 +....+....+....+....+....+....+....+....+....+.PKEEEEEEE
      0   .1   .2   .3   .4   .5   .6   .7   .8   .9  1.0
THE ABSCISSA IS     ISWP.  MIN= 1.00E+00 MAX= 1.50E+02
```

Figure 10.7. Residuals against sweep (bias = 20)

184

the residuals are plotted as the ordinate, with the values being scaled to fit between 0 and 1. Figure 10.6 shows that the spot values become constant with sweep number. We can see that the v-component of velocity and the pressure both fall then rise to a steady value, ε rises then falls and both k and the w-component rise, fall and then rise again. From the printed numerical values in the RESULT file, the variation in all the variables during the final ten sweeps occurs in the third or fourth significant figure. Looking at Figure 10.7, the residuals can be seen to fall steadily with a small departure for the residual of the w-component equation near the end of the run.

Using the spot values and residuals as a guide, we can see that the solution achieved after 150 sweeps is converged to an accuracy of three significant figures. No further running of the solver is required and so we can turn our attention to the results calculated throughout the flow field. PHOTON, the post-processor of PHOENICS, can read the data produced by EARTH and Figures 10.8 and 10.9 show the velocity vectors of the simulated flow for two different views. These diagrams have been produced using commands similar to those given for the PHOTON picture in the first example, see Section 10.3.6, and it is interesting to note that PHOTON has taken account of the fact that there is no flow within the vehicle surface and has not plotted any data there, and that the vectors appear at the cell positions where they were calculated. Some graphics programs produce data on a different mesh to that used by the calculation. This can be useful if the mesh is unstructured. In Figure 10.8 the velocity field is displayed for the area of the rearscreen of the vehicle. The flow is seen to separate from the rearscreen about two-thirds of the way down the screen and a small vortex is found in the screen–boot intersection. In reality, a car of this shape would have a much larger area of separated flow over the rearscreen and so the results we have obtained, although numerically converged, do not quite agree with what we might expect. Looking at Figure 10.9, we can see the flow field over the roof of the vehicle. Here, the boundary layer is hardly noticeable, with nearly all the variation taking place in the first two cells from the surface. All this suggests that the mesh is not refined enough near the vehicle surface.

Refining the computer model is straightforward, now that we have the tools to build the mesh, and a model Q1 file already exists. A second mesh has, therefore, been created using the same distribution of cells along the vehicle surface and in the flow field, but the biasing

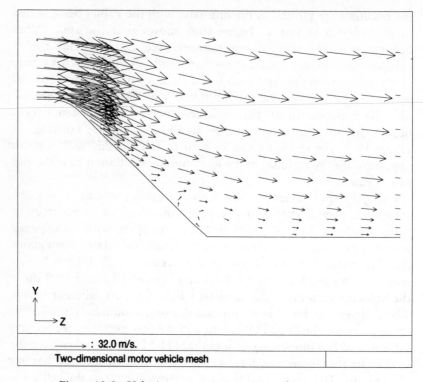

Y
↑
└──→Z

──────→ : 32.0 m/s.

Two-dimensional motor vehicle mesh

Figure 10.8. Velocity vectors at rearscreen (bias = 20)

parameter has been increased from twenty to fifty for the blocks
ahead of, above and behind the vehicle. As we have not changed the
structure of the mesh, or the cell numbers in each of the local mesh
directions, no changes are required to the Q1 file, but SATELLITE
has to be run again, before running EARTH, so that the new file
containing the cell corner points is read and new geometry data is
written for EARTH.

For this second mesh, a run of 150 sweeps has been made by
EARTH, and the RESULT file shows very similar trends to those
found with the original mesh for the variation of the spot values at
the monitor location and the residuals with sweep number. Using
PHOTON to look at the velocity field, Figures 10.10, 10.11 and
10.12, several interesting features can be seen. By changing the
biasing parameter in the mesh building process, more cells are placed
near the vehicle surface and so there are more vectors near the surface. This leads to a solution which has a larger area of flow separa-

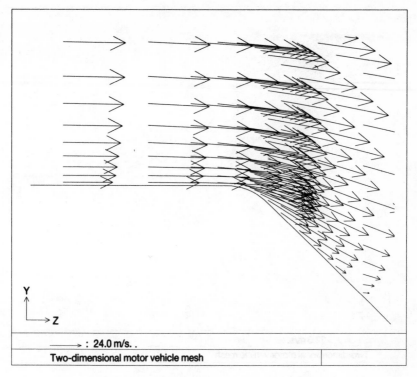

: 24.0 m/s. .

Two-dimensional motor vehicle mesh

Figure 10.9. Velocity vectors on roof (bias = 20)

tion (Figure 10.10) and a much better definition of the boundary layer on the surface (Figure 10.11). Consequently, the overall flow picture near the car, Figure 10.12, can be seen to be qualitatively correct. Note that in this last diagram the flow slows down as it approaches the front of the vehicle and that it speeds up at the front of the bonnet and at the vehicle roof where the surface changes direction rapidly in space. Also, as well as the separation at the rearscreen, there is a region of separated flow behind the vehicle where two vortices can be seen. All of these features can be seen when the physical vehicle model is placed in an airflow in a wind tunnel.

This refinement process can be continued and the biasing parameter continually increased. A case has been run with the parameter set to 200 above the vehicle and 50 ahead of it and behind it. One-hundred-and-fifty sweeps have been calculated yet again. Figure 10.13 shows the residual variation with sweep number and,

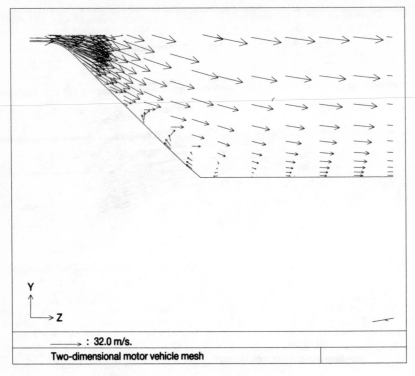

Figure 10.10. Velocity vectors at rearscreen (bias = 50)

from this, the variation of the residual error for the w-component of velocity can be seen to oscillate wildly. There is also a smaller oscillation of the pressure residual. This shows that the solution is not progressing satisfactorily. Confirmation of this is found by looking at the spot values which also oscillate in magnitude when plotted against sweep number. One way of suppressing this oscillation is to run the solution with more relaxation by using smaller relaxation factors, and another way is to create a mesh of the domain which is smoother.

At this stage in the process we appear to have reached the limit of accuracy with this particular distribution of cells. Whilst further simulations could be produced, we will leave this two-dimensional calculation knowing that we have found a solution which is qualitatively correct for the velocity field.

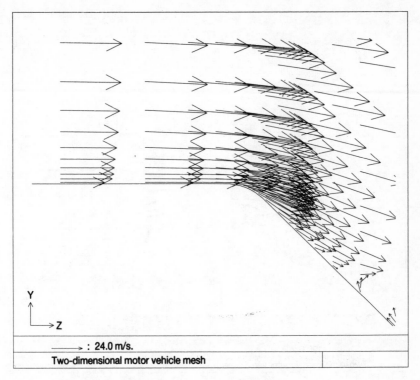

Figure 10.11. Velocity vectors on roof (bias = 50)

10.4.5 *A note on three-dimensional calculations*

Knowing that a simulation is qualitatively correct is often all that is required of a simulation. Such simulations can provide an engineer with sufficient information to make sensible choices about the design of an object and the effect of these choices on the flow of the fluid. In the case of the flow about a car, however, engineers must know something about the forces and moments produced on a vehicle by the flow. This is quantitative information.

If we carry out CFD simulations of the flow about a car, we will also want to know what the forces and moments are that the simulated airflow would produce. Several manufacturers of vehicles and CFD software authors have performed such simulations in three dimensions (Shaw, 1988; Shaw and Simcox, 1988; Hawkins *et al.*,

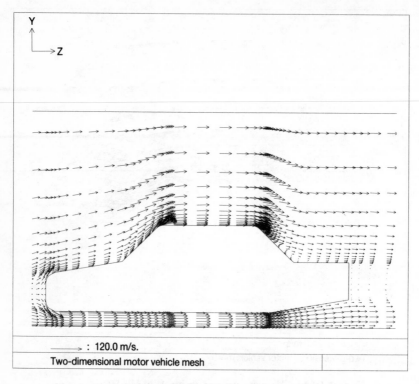

Figure 10.12. Velocity field around car (bias = 50)

1990), but the results have not been promising. These simulations have shown that the qualitative picture of the flow produced by the simulation is in good agreement with that found in wind tunnel tests. Also there is a good agreement between the prediction of fluid pressure on the surface of the vehicle and that found by experiment. So far so good, but the bad news starts when the predicted pressures are integrated over the vehicle surface, for each cell face on the surface, to give a measure of the forces and moments on the vehicle body. Even if we take the problems of the modelling of the wheels of the vehicle and the drag due to viscous shear into account, the predicted drag is in poor agreement with the experimental values.

One source of the error between the predicted forces and moments and the experimental values comes from the integration process itself. A vehicle in a real flow sees what is in effect an infinite number of fluid particles over the vehicle surface, giving a pressure which

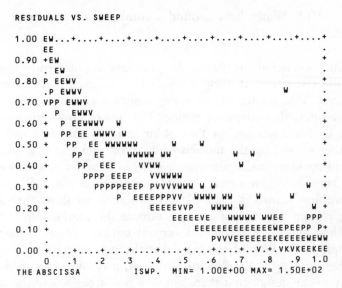

```
RESIDUALS VS. SWEEP

1.00 EW...+....+....+....+....+....+....+....+....+....+....+
     EE                                                       .
0.90 +EW                                                      +
     . EW                                                     .
0.80 P EEWV                                                   +
     .P EWWV                                          W       .
0.70 VPP EWWV                                                 +
     . P  EWWV                                                .
0.60 +  P EEWWWV  W                                           +
     W  PP EE WWWV W                                          .
0.50 +   PP  EE WWWWWW     W    W                             +
     .    PP  EE   WWWWW WW              W  W                 .
0.40 +    PP  EEE    VVWW              W                      +
     .    PPPP EEEP  VVWWWW                                   .
0.30 +    PPPPPEEEP PVVVVWWW W W                    W  +
     .       P  EEEEPPPVV  WWWW WW     W     W      .
0.20 +           EEEEEVVP  WWWW W                           +
     .           EEEEEVE   WWWWW WWEE      PPP
0.10 +              EEEEEEEEEEEEEEEEWEPEEPP P+
     .              PVVVEEEEEEEEKEEEEEWEWW
0.00 +....+....+....+....+....+....+....+....+..V.+.VKVKEEKEE
     0   .1   .2   .3   .4   .5   .6   .7   .8   .9  1.0
THE ABSCISSA        ISWP.  MIN= 1.00E+00 MAX= 1.50E+02
```

Figure 10.13. Residuals against sweep (roof bias = 200)

varies continuously over the vehicle surface. When several hundreds of thousands of cells are used in the simulation, the cost of the computer time alone for the simulation is greater than the cost of the corresponding physical experiment, and the number of cells on the surface might still be only of the order of a few thousand. This means that the simulation cannot capture the same level of variation that the vehicle in a physical experiment would and, consequently, the numerical integration is very inaccurate. This would still be the case even if the values of the pressure at all the mesh points on the surface were exact.

I mention this problem to give the reader something to think about. The aim of CFD in engineering is to produce results which are useful in the design process, not to produce pretty colour pictures for the office wall or for your manager. This means that companies and individuals must decide whether CFD is the right tool for their particular application. The ways of doing this are discussed further in Chapter 12. It should be mentioned here that, for some problems, CFD might be the only means of analysis and it might also be cheaper than the experiment. The next example shows a problem that is well suited to CFD, giving real insight into the technical problem.

191

10.5 Water flow around a combustion chamber

10.5.1 *Producing a specification*

In many industrial problems, the geometry is sufficiently complex that the restriction of using a regularly structured mesh cannot be tolerated. One source of extremely complex geometry is an automotive internal combustion engine. Two major flow situations that occur in this device are the flow of air and fuel into the combustion chamber caused by the motion of a piston and the flow of water around passages inside the engine where the water removes excess heat from the engine casing. In this final case study we will consider the problem shown in Figure 10.14, where water flows through an inlet, around the cooling passages outside the combustion chamber and then flows out through a vertical outlet. This is a simplified example of the flow of cooling water through an engine.

When looking at the flow of a coolant, it is important to the efficiency of the design that there are very few areas where the flow is separated. In such regions, the fluid moves slowly over the hot surfaces and so the heat cannot be removed from these surfaces in an efficient way. One objective of a CFD analysis of such a situation is to determine where it is, within the cooling system, that these areas of separated flow occur, if they do occur. Then modifications can be made to the geometry of the internal passages to ensure that such areas do not occur or that their existence is minimized. Further, it is possible that CFD can give a good estimate of the pressure loss in the fluid as it passes through the system and this can be used to specify the required pressure head of the water pump.

From Figure 10.14 we can see that the bounding surfaces of the geometry are simple planes or cylinders and so the production of a mesh should not be difficult. When considering the flow through the system, only three types of boundary can be present. These are an inlet, an outlet and a series of solid walls. At the inlet, the velocity is 5 m/s and the width is 0.0232 m. For water at 15 °C, the Reynolds number based on the inlet width is

$$Re = \frac{\rho V_{inlet} D}{\mu} = \frac{(1000)(5)(0.0232)}{11.4 \times 10^{-7}} = 1.018 \times 10^5 \qquad (10.16)$$

As the height of the inlet is 0.02 m, the Reynolds number based on height will be much the same. Given these values of the Reynolds number, the flow can be assumed to be turbulent. This can be determined by considering the flow in a pipe (see Duncan *et al.*, 1970,

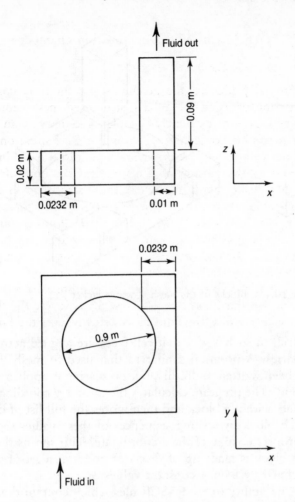

Figure 10.14. Geometry of water jacket

Chapter 7), where the critical Reynolds number for the flow to change from a laminar flow to a turbulent flow, the transition process, is about 2000.

10.5.2 *Producing a mesh*

To produce a mesh for this problem, we can split the geometry into a series of blocks, as shown in Figure 10.15. Then a simple structured mesh can be built in each block and the blocks connected at

193

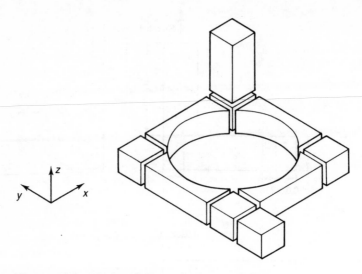

Figure 10.15. Blocks of the mesh – water jacket

their boundaries in such a way that cell faces are aligned across the block boundaries. A program similar to that used to mesh the car example has been written to do this, but no biasing is applied to the cell distribution. The program calculates the x- and y-coordinates of the mesh points within a block and then writes the full list of coordinates for each block by writing sequences of these values together with the appropriate value of the z-coordinate. This means that the mesh in each block is made up of sheets of nodes at a set of planes which are defined by having constant values of z.

STAR has the ability to read ASCII files which contain the mesh data. Two files are required to specify the mesh: one file contains the list of the x-, y- and z-coordinates for each point in the mesh, known to STAR as a *vertex*; and the other file contains a list of the identification numbers of the vertices that are connected to each element. These files are written directly by the mesh generation program.

To calculate the mesh, the program must be given a set of parameters that state the numbers of cells in the various blocks. These parameters are labelled $n1$, $n2$, nz, $n$inlet and $n$riser, as shown in Figure 10.16. A complete mesh is illustrated in Figure 10.17 for values of these parameters set to 7, 7, 8, 10 and 20 respectively. This is the mesh that we will use.

194

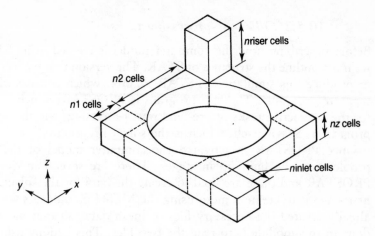

Figure 10.16. Mesh layout parameters for water jacket

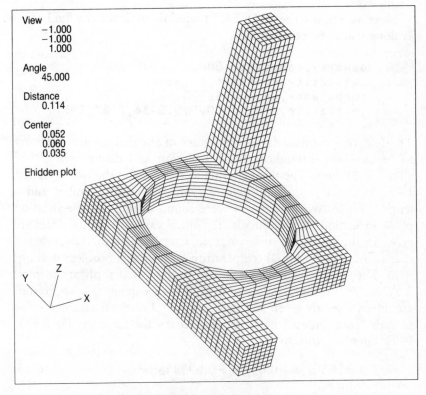

Figure 10.17. Mesh of the water jacket

195

10.5.3 *Other pre-processing tasks*

Before describing how the computer model is created using STAR, we must outline the structure of STAR. The version that we have run to produce this simulation is STAR v2.004, which consists of two separate programs. The first program is PROSTAR, which is used for the interactive tasks of pre- and post-processing, and the second program is STAR itself, which is the solver program.

Once PROSTAR is activated the computer model of the flow problem can be built up in stages. There are several modules to PROSTAR and these are used to create the data for the solver. The first stage is to create a mesh using the MESH module. As we have already created the necessary files of mesh data, all that has to be done in this module is to read the two files. This is done using the VREAD and CREAD commands, where the V refers to vertices, which are the corner points of the cells, and the C refers to the cells themselves.

Next we can use the PROPERTY module to define the fluid. This is done using the commands:

```
density,constant,1000.
lviscosity,constant,11.4e-4
turbulence,ke,1.018e5,0.02
initialize,0.0,5.0,0.0,0.0938,2.87,293.
```

The first two commands set the values of the density and viscosity to be constant throughout the calculation and define appropriate values in SI units. The third command switches on the two-equation $k-\varepsilon$ turbulence model and gives a typical Reynolds number and a length. These two parameters were found during the specification phase in Section 10.5.1. Finally, the initial values of the variables are given in the following order: $u, v, w, k, \varepsilon, T$. Here the last value T that is listed is the initial temperature and for this problem it is not used. The velocity values are taken to be those that apply at the inlet. To calculate the initial values of the turbulence quantities, approximate inlet values are given and these are found from an assumed value of turbulence intensity of 5 per cent. Using the formulae for turbulence intensity, this gives

$$k = \tfrac{3}{2} I^2 V_{\text{inlet}}^2 = \tfrac{3}{2}(0.05)^2 5^2 = 0.0938 \text{ m}^2/\text{s}^2 \tag{10.17}$$

An approximate value of the mixing length is known for a flow near a wall, that is

$$l = \frac{\varkappa y}{c_\mu^{0.75}} = \frac{(0.41)(0.01)}{(0.09^{0.75})} = 0.024 \text{ m} \qquad (10.18)$$

Here $\varkappa$ is a constant for a boundary layer and the value of y is taken to be half the inlet height. The value of the mixing length that is derived will be a maximum value and so an average value will be used for the mixing length of 0.01 m. Finally, the value of ε is found from the additional turbulent viscosity. This is calculated from equation (2.17) as

$$\nu_T = c_\mu k^{0.5} l = (0.09)(0.0938^{0.5})(0.01) = 2.76 \times 10^{-4} \text{ m}^2/\text{s} \qquad (10.19)$$

and so, from equation (2.18), ε is given by

$$\varepsilon = \frac{0.09 k^2}{\nu_T} = \frac{(0.09)(0.0938^2)}{2.76 \times 10^{-4}} = 2.87 \text{ m}^2/\text{s}^3 \qquad (10.20)$$

Once the fluid properties have been defined, the boundary conditions have to be set using the BOUNDARY module. STAR assumes that any unspecified boundary is a solid wall and so this simplifies the specification of the boundary conditions considerably. All we have to do is specify the location of the inlet and the outlet and then define the conditions that apply at these two boundaries. The surface of the cells of the mesh can be plotted on the screen in PROSTAR. This is done using the PLTYPE,QHIDDEN command which displays a simplified hidden-line plot. Then the cursor can be used to pick the cell faces that are at the inlet and the outlet. The commands used to do this are

```
bcross,add,1
bcross,add,2
```

These define one set of faces to be region 1 and the other set of faces to be region 2. These regions are then associated with the boundary conditions using the commands

```
rdefine,1,inlet
0.0,5.0,0.0,1000.,0.0938,2.87
rdefine,2,outlet
```

These specify that the first region is an inlet at which the values of $u, v, w, \rho, k, \varepsilon$ are given as listed. Similarly, the cell faces in region 2 are defined to be the outlet.

The last module to be used is the CONTROL module, where the data to control the numerical solution is provided. Within this module, the commands are used to control the initial run of the solver. They are listed below, together with the variations that would be used to carry out a restart calculation. The commands are (with the restart commands given in square brackets):

```
time,0.005,steady
iter,10,500,0.001 [iter,100,500,0.001]
simple,on
rdata,none      [rdata,restart,binary]
wdata,post,binary
relax,0.1,0.1,0.1
monitor,101
```

These specify that the calculation is a steady state process: that ten iterations out of a maximum number of 500 are to be run with the program stopping if the residual falls below 0.001; that the SIMPLE algorithm is to be used; that no initial data is to be read from a file but that a restart calculation would read in this data; that a file suitable for post-processing is to be written in binary format; that the linear relaxation factors for the pressure, velocity components and the turbulence parameters, respectively, are set to 0.1 and that the variables in cell 101 are to be printed at every iteration.

Finally, the data necessary for the STAR solver program is written using the commands

```
geomwrite,8
probwrite,10
```

which write the geometry data to the file numbered 8 and all the other data to the file numbered 10.

10.5.4 *Running the solution*

At first, ten iterations are run to check that the model is working. The STAR solver produces output which lists the residuals and the monitored values at each iteration, and these show that the residuals are decreasing except for ε which is increasing slightly. This rise in the ε residual is not too much of a problem, but in an attempt to get all the residuals reducing, the relaxation factor for the turbulence variables has been reduced to 0.01. Seventy iterations have then been run starting from the initial values again.

Yet again, all the residuals have decreased except for those from the ε equation, but the rate of increase of this is clearly reducing. This suggests that if further iterations are run with the same relaxation factors the ε residual should start to reduce. At the monitor location the velocity components and pressure are changing rapidly, but the turbulence parameters are only moving slowly due to the severe relaxation. To continue the solution, 200 further iterations have been run with the same relaxation factors. The printout now shows that all the residuals are falling and that the values of the variables at the monitor locations are changing less and less.

Finally a further eighty iterations have been run with the relaxation factor for pressure kept the same at 0.1, but the relaxation factor for the velocity components has been increased to 0.4 and for the turbulence variables it has been increased to 0.1. Initially, all the residuals have fallen but later the ones for the velocity components and pressure have started to oscillate. The residuals for the velocity components have decreased by factors of several hundred and for pressure by several thousand. By looking at the values of the variables at the monitor location we can see that they are now only changing in the third significant figure every sixteen iterations and so, effectively, the solution is converged.

10.5.5 *Analysing the results*

To look at the results graphically, we can use PROSTAR again. First of all we must tell PROSTAR to access the computer model that we set up in PROSTAR during the pre-processing phase (file 16) and the file of results that was created during the solution phase (file 9). This is done using the commands:

```
resume,16
load,9
```

Once PROSTAR has read the data that it needs for post-processing, various commands can be used to plot the data. For this example it is likely that the most useful information will come from a plot of the velocity vectors calculated by STAR. These will allow an engineer to make a qualitative assessment of the way in which the flow is behaving. For example, the commands:

```
vescale,0.5
poption,vector
getcell,all,none
plty,section
surf,on
edge,off
view,0,0,1
spoint,0.0,0.0,0.01
cplot
```

will plot the picture shown in Figure 10.18. This is a velocity vector plot at half the height of the main flow channel. The commands given above are used to scale the vectors to a reasonable size, set the plot type to vector, use all the available cells, produce a section plot on a plane through a point (0.0,0.0,0.01) together with the surfaces taken as if viewed from a point on the z-axis. The last command actually plots the picture.

From this diagram we can see that the flow comes in through the inlet and then splits into two to go around the cavity formed by the combustion chamber. At the point of splitting, the magnitude of the vectors is small, as it is in the upper left and lower right corners of the flow system. Also, where the two streams come together, the flow velocity is small. If the surfaces of the cavity were hot, and the water was being used to transport heat away from these surfaces, then the

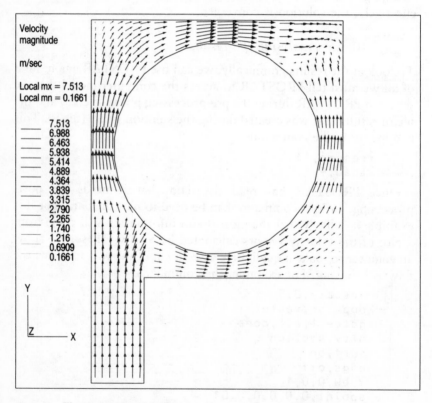

Figure 10.18. Velocity vectors around water jacket

heat transfer would not be very good in these areas. Figure 10.19 shows a similar plot through a vertical section which passes through the outlet channel. This shows that the flow separates from the passage surfaces at the entry to the outlet passage and we can see that there is a vortex near to the left hand wall of the channel where the flow velocity is very small. This area of separated flow restricts the effective width of the channel, and leads to pressure losses. The outlet passage could be redesigned to remove this separation region, reducing the pressure losses in the system and also improving any heat transfer in the area of the start of the outlet channel.

Even in large systems of pipes, it is this sort of information that can be used to improve the fluid flow within a system and have a beneficial impact on engineering design.

One thing that is clear from these two diagrams is that the

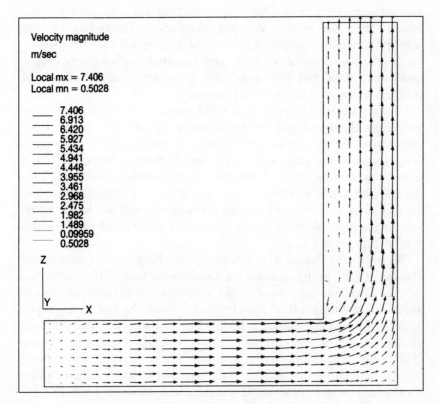

Figure 10.19. Velocity vectors in vertical plane through outlet

boundary layers near the walls of the passages have not been modelled very accurately. This may not affect the qualitative nature of the prediction, but it will affect any quantitative data such as the variation of pressure through the domain. If this simulation is to be improved it would have to be remodelled using a mesh with more cells and, perhaps, with the cells being biased towards the walls of the passages.

10.6 A review of the usefulness of CFD

From this set of three examples we can find pointers to the usefulness of CFD. The first example shows that simple laminar flows can be calculated to a high degree of accuracy with little effort. The other two examples show that we have to be careful in using CFD if the flow is more complex. With the example of the flow over a car, the predicted data provides a reasonable simulation both qualitatively and quantitatively, but when the numerical pressure data is integrated, then the results are very inaccurate. Depending on the information that is required this could be a good simulation or it could be a poor simulation. In the third example we have only sought qualitative data and this simulation provides a large amount of information that is of use to an engineer.

Looking at other examples of CFD, the use of the technology in predicting the weather is extremely useful and accurate in most circumstances. However, the simulations are restricted in that the mesh size cannot be too great as the calculations have to be performed in a reasonable time and not use too much computer memory. This means that freak weather events which have a spatial dimension smaller than the distance between mesh points will not be predicted very accurately. Without infinite computer power and memory this will always be the case.

We have also discussed, in Chapter 1, the King's Cross inquiry and the use of CFD in determining the cause of the fireball that occurred. The CFD calculations pointed out a possible mechanism for this in the form of the so-called *trench effect*, where the hot gases stayed near the floor of the escalator tunnel. This phenomenon had not been thought of before and so experiments were carried out to investigate if it could actually occur in practice. These experiments showed similar flow patterns, confirming that the mechanism predicted using CFD could occur in practice. It was the combination of both the CFD prediction and the subsequent experimentation that made this study so conclusive. It is becoming more common that these two predictive techniques, experimentation and computation, have to be used together. They should be seen as complementary means of carrying out investigations, not as opposing strategies.

In summary, it can be said that CFD does have its uses but that the results of simulations need to be considered carefully before they are used.

11 Modelling flows with additional complexity

So far, we have considered the ways in which CFD tools can be used to predict flows which can be classified as incompressible and viscous. Many industrial flow problems encountered outside the aircraft industry can be described as flows that fall into this category. This means that many flows can be modelled by applying the techniques that we have discussed. If we are to model some, or all, of the other categories of flows, we must determine the modifications that need to be made to our modelling technique. In particular, the modelling of four additional features will enable a large proportion of the flows that are not simply incompressible and viscous to be modelled. These four features are as follows:

- *The prediction of heat transfer within a flow*. In this case an additional equation, the energy equation, which describes the transport of heat energy through a fluid, has to be solved.
- *The effects of compressibility*. Many fluids in motion exhibit the effects of compressibility. This occurs when the density of the fluid changes in the flow field.
- *The existence of multiple phases within the flow*. In some flow problems two or more fluids can flow together. For example, a liquid and a gas could move together. Also, the transport of solid material in a fluid can be described as being a multi-phase flow.
- *The inclusion of combustion*. When a fuel is burned chemical changes take place and energy is released. This can occur in a fluid that is already flowing or it can cause a fluid to flow.

In this chapter we will discuss, in a simple way, each of these topics in turn. It should be noted that the material that follows is not meant

204

to be an exhaustive treatment. The aim of this chapter is to highlight some of the modifications that are made to the modelling process which enable these features to be catered for. This will help the analyst to look in the right places for further information, if the need to carry out such modelling ever arises.

11.1 Modelling flows with heat transfer

11.1.1 *The effects of heat transfer on a flow*

Heat transfer is the movement of internal energy around a system. It can occur in three main ways: *conduction*, where the agitation of molecules transfers the energy from one molecule to another; *convection*, where the transport of material transfers the energy from one place to another, and *radiation*, where electromagnetic fields are the mechanism of energy transfer. The textbook by Rogers and Mayhew (1980) provides a good basic introduction to the subject.

Within a given situation, all three modes of heat transfer might occur. For example, heat might flow through a solid by conduction and then be transferred into a fluid where it is convected away with the fluid, and if, say, flames are present they will radiate heat energy all around. However, in the context of fluid flow, it is convection that is the most important and so we will concentrate on this mode of heat transfer.

There are two main types of convection. Let us consider the situation where a fluid is forced by some pressure forces to flow over a hot object. Some of the heat is removed from the hot object and convected away. This is known as *forced convection*. Conversely, a hot object might heat the surrounding stationary fluid, causing its density to reduce locally. When this happens the hotter fluid rises through the colder fluid, an effect of gravity, and we have what is known as *natural convection*. In the first case, the flow takes place and the heat transfer is a secondary effect, whereas in the second case, the heat transfer actually drives the flow of the fluid.

When we started our discussions of the modelling of fluid flow, we had to derive mathematical relationships between the variables that could then be converted into numerical equations. Similarly, when modelling heat transfer by convection, we have to have some mathematical model of the energy transfer process and so now we must look at how this can be derived.

11.1.2 *The energy equation for heat transfer*

When modelling incompressible, viscous flows we must use the momentum and continuity equations to calculate the velocity components and the static pressure of the fluid. If we are to model heat transfer by convection, then we must also find some relationship between the flow variables and a property related to the heat flow in the fluid. The property we normally choose to do this is the temperature, which we need to calculate throughout the flow domain. This is done by using an energy equation derived from the first law of thermodynamics. The derivation is shown in detail by Schlichting (1979) and in an abbreviated form by Chapman (1984).

If we consider a given patch of fluid, as we did in Chapter 2, the first law states that the heat entering the patch can lead to some combination of two effects. It can raise the internal energy of the fluid in the patch and it can enable the patch of fluid to do work on its surroundings. Hence, by considering the rates at which these events occur, we can write the first law of thermodynamics as

$$\frac{dQ}{dt} = \frac{dE}{dt} + \frac{dW}{dt} \tag{11.1}$$

where Q is the heat energy entering the patch by conduction, E is the internal energy of the patch and W is the work done by the fluid in the patch.

This can be rewritten in terms of the heat added per unit volume q, the work done per unit volume w and the internal energy per unit mass, the specific internal energy e, i.e.

$$\frac{dq}{dt} = \rho \frac{de}{dt} + \frac{dw}{dt} \tag{11.2}$$

where ρ is the fluid density.

Dealing with each of these terms, the first is the rate at which heat energy is conducted into the patch, which can for two dimensions be shown to be given by

$$\frac{dq}{dt} = k \left(\frac{\partial^2 T}{\partial x^2} + \frac{\partial^2 T}{\partial y^2} \right) \tag{11.3}$$

where k is the constant thermal conductivity of the fluid.

The second term contains the rate of change of specific internal energy in the patch, and it is possible to describe this as the product of the specific heat at constant pressure c_p and the rate of change of the fluid temperature T, which in this case is the substantive derivative (see Section 2.2.1) of the temperature. In two dimensions this gives

$$\rho \frac{de}{dt} = \rho c_p \frac{dT}{dt} = \rho c_p \left(\frac{\partial T}{\partial t} + u \frac{\partial T}{\partial x} + v \frac{\partial T}{\partial y} \right) \tag{11.4}$$

The third term is the rate of work done per unit volume of fluid and this can be taken as being due to viscous forces alone if the fluid is incompressible. This means that

$$\frac{dw}{dt} = -\mu \left[2 \left(\frac{\partial u}{\partial x} \right)^2 + 2 \left(\frac{\partial v}{\partial y} \right)^2 + \left(\frac{\partial u}{\partial y} + \frac{\partial v}{\partial x} \right) \right] \tag{11.5}$$

Here the negative sign shows that this is, in fact, work done on the fluid in the patch, not the work done by this fluid. Combining these terms together we obtain the equation for the transport of temperature through the domain, which is

$$\left(\frac{\partial T}{\partial t} + u \frac{\partial T}{\partial x} + v \frac{\partial T}{\partial y} \right) = \frac{k}{\rho c_p} \left(\frac{\partial^2 T}{\partial x^2} + \frac{\partial^2 T}{\partial y^2} \right)$$
$$+ \frac{\mu}{\rho c_p} \left[2 \left(\frac{\partial u}{\partial x} \right)^2 + 2 \left(\frac{\partial v}{\partial y} \right)^2 + \left(\frac{\partial u}{\partial y} + \frac{\partial v}{\partial x} \right) \right] \tag{11.6}$$

This equation is very similar in form to the momentum equations for laminar flow that we discussed in Chapter 2. It is a non-linear equation describing the temperature in the fluid as being related to the flow velocity and some properties of the fluid. To produce a numerical analogue of this equation, any of the techniques described in Chapter 3 can be used and the solution of the equation can be inserted into the full solution process. If the SIMPLE algorithm is used to calculate the variables, the temperature is found after the pressure equation has been used to update the static pressure and to correct the velocity components that have been derived from the momentum equations.

Such a procedure will be valid for laminar flows where the effects of any changes in fluid density can be ignored. This would be the case in forced convection problems. The effects of turbulence on the situation and of any changes in the density, the so-called buoyancy effect, still have to be taken into account.

As a last note on the energy equation, we can see from equation (11.6) that the boundary conditions are likely to be the specification of the temperature on a boundary or the specification of the normal derivative of temperature. In some circumstances these will be known directly, but often any derivatives will have to be found from empirical data.

11.1.3 *The effects of turbulence on heat transfer*

In Sections 2.2.2 and 2.2.3, we looked at how to account for the fact that in a turbulent flow the velocity components can be thought of as being made up of a mean value and a fluctuating component. We saw there that such an analysis produced additional terms in the momentum equation which can be modelled as additional stresses. If we carry out a similar procedure for the energy equation (11.6), then further heat flux terms are generated. These again have to be modelled by some means.

As it can be shown that there are analogies between the modifications to the momentum equations due to turbulence and the modifications to the energy equation due to turbulence, simple ways of modelling the additional heat flux terms are often used.

11.1.4 *Buoyancy effects*

When a fluid is heated its density changes. This means that as the density changes so gravity will exert a different force on a patch of fluid. Consequently, hot fluids will try to rise through cold ones. This is the mechanism of what is known as *natural* or *free convection*, and this has to be modelled in some CFD simulations. Take, for example, the case of a fluid inside a double-glazing system, Figure 11.1. There, the left hand wall of the cavity is hot, the right hand wall is cold and the other two walls have no heat flowing through them. Gravity can be seen to act vertically downwards, parallel to the hot and cold walls. Due to the temperature of the walls, on the left hand side the fluid will be hotter than on the right hand side, and the density of the fluid will be lower on the left than on the right. As gravity will exert a lower force on the fluid at the left, there will be a movement of fluid up the hot wall and down the cold wall. To model such situations the addition of the energy equation is not sufficient, as we have to include the effect of gravity in the momentum equations. This is done by adding an effective force term due to the density variation to the right hand side of the momentum equations. This force acts in a direction parallel to the direction of the gravity force.

Looking at Figure 11.1, where the problem is shown with the y-axis being vertical, i.e. gravity acts in the negative y-direction, we

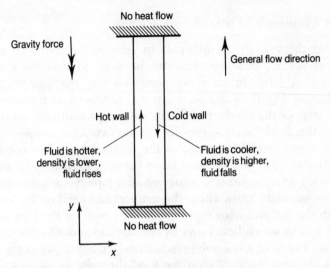

Figure 11.1. The double-glazing problem

can see that the density changes will lead to an additional term X in the y-momentum equation. This term can be modelled, for our usual patch of fluid, as

$$X = \frac{1}{\rho} \, g(\rho_f - \rho)\delta x \delta y \qquad (11.7)$$

where ρ is the local density of fluid and ρ_f is a reference density. Hence, when the density ρ is less than the reference ρ_f the fluid will have a positive force on it. This density relationship can be converted into a temperature relationship by using the coefficient of volume expansion β to give

$$X = g\beta(T - T_f)\delta x \delta y \qquad (11.8)$$

Here, T_f is a reference temperature and β is defined by

$$\beta = \frac{1}{\mathbf{v}} \left(\frac{\partial \mathbf{v}}{\partial T}\right)_p = -\frac{1}{\rho} \left(\frac{\partial \rho}{\partial T}\right)_p \qquad (11.9)$$

where $\mathbf{v}$ is the specific volume of the fluid and the p denotes that the derivative is calculated at constant pressure. Combining the equation for X, equation (11.8), with the momentum equation in the y-direction, equation (2.9), gives

$$\frac{\partial v}{\partial t} + u \frac{\partial v}{\partial x} + v \frac{\partial v}{\partial y} = -\frac{1}{\rho} \frac{\partial p}{\partial y} + \frac{\mu}{\rho} \left(\frac{\partial^2 v}{\partial x^2} + \frac{\partial^2 v}{\partial y^2}\right) + g\beta(T - T_f) \qquad (11.10)$$

It is the additional term that is known as the *buoyancy* term.

209

11.1.5 *Conjugate heat transfer problems*

If we set the velocity components to zero in the energy equation, equation (11.6), this equation can be used to describe the heat transfer in a solid. In some problems, such as the flow in a thick walled pipe, Figure 11.2, we might be interested in modelling the total system of the conduction in the solid pipe wall and the convection in the fluid. Such a problem is known as a *conjugate heat transfer problem* and has some of the thermal boundary conditions set at solid boundaries, not fluid boundaries. Considering Figure 11.2 as showing an axisymmetric situation, the pipe has a hot outer wall at which we might know either the temperature itself or the heat flux through the surface. Moving towards the centre of the pipe there is a solid wall in which heat flows by conduction and all velocity terms are zero. Then there is a solid–fluid interface before we come to the fluid in the pipe itself. Within the fluid the velocity increases away from this interface to a maximum at the centre of the pipe, and the temperature falls towards the centre.

Many CFD software packages can solve such problems, but there can be numerical difficulties at the fluid–solid interface. These problems can arise from the distribution of cells with some finite volume programs, where a cell might straddle the interface. This would not happen with finite element programs. Also, the modelling of conjugate heat transfer with turbulent flows can lead to problems when any of the discretization techniques are used, as log-law profiles have to be applied at the interface.

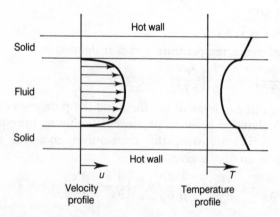

Figure 11.2. A conjugate heat transfer problem

11.1.6 *Some non-dimensional groups*

When we looked at defining incompressible viscous flows, we saw that a useful parameter in classifying the flow is the Reynolds number. There are several non-dimensional groups that are useful when considering heat transfer problems and these are:

- *Prandtl number*, which is defined as

$$Pr = \frac{\mu c_{\mathrm{p}}}{k} \tag{11.11}$$

and can be seen to be the ratio of viscous diffusion of momentum to thermal diffusion through conduction. Typical values of Pr for gases are in the range 0.65 to 1.0, with air having a value about 0.7. By comparison water has a value of about 6.0 at room temperature.

- *Nusselt number*, which is defined as

$$Nu = \frac{hd}{k} \tag{11.12}$$

where d is a typical length and h is the heat transfer coefficient defined as the surface flux of heat $\dot{q}$ divided by some temperature difference, i.e.

$$h = \frac{\dot{q}}{T_{\mathrm{s}} - T_{\mathrm{f}}} \tag{11.13}$$

where T_{s} is the temperature of the surface and T_{f} is a reference temperature, say of the fluid surrounding the surface. Nusselt number is a non-dimensional measure of the heat transfer through a surface.

- *Grashof number*, which is defined as

$$Gr = \frac{gd^3 \beta \, \Delta T}{\nu^2} \tag{11.14}$$

where g is the acceleration due to gravity, d is a typical length, β is the coefficient of volume expansion, ΔT is a temperature difference and ν is the kinematic viscosity. This parameter is used to characterize natural convection problems.

211

11.2 Modelling flows that are compressible

11.2.1 *Some features found in compressible flows*

Flows that are compressible have a varying density of the fluid throughout the flowfield. These flows exhibit some features that are not found in incompressible flows. Amongst these are the discontinuities known as shock waves, where fluid variables change rapidly over a small spatial distance. Many books show pictures of the types of flow that can be found (see van Dyke, 1982). These features are found in addition to the features of viscous flows already discussed.

One way of classifying a compressible flow is by the parameter known as the Mach number. This is defined as the ratio of local flow speed V_{local} to the local speed of sound in the fluid a, or

$$Ma = \frac{V_{local}}{a} = \left(\frac{u^2 + v^2 + w^2}{a^2}\right)^{1/2} \tag{11.15}$$

From this we can see that once a flow is moving, the Mach number is not zero. Standard texts on compressible flow, such as Shapiro (1953), show that if the Mach number is small, say less than 0.2, then the flow may be considered incompressible, but when the Mach number is greater than this, then the flow must be considered as compressible. If the local Mach number is less than unity everywhere, *a subsonic flow*, shock waves will not appear, and the flow will qualitatively behave like an incompressible flow, hence our modelling technique need hardly be altered; whereas, if the flow has regions where the Mach number is greater than one, which are known as *supersonic flow regions*, then shock waves can appear. If we consider the flow around an object, the Mach number well away from the object may be very small; however, the flow must accelerate around or through the object, and so the local Mach number can be much greater in some places. For example, the flow through a narrow gap such as that between an aircraft wing and a slat can reach supersonic speeds if the Mach number of the free stream, that is, away from the wing, is as small as 0.2.

The change in the observed flow types for supersonic flow areas suggests that something fundamental must be happening in the flow that is different from what happens in incompressible flow. By looking at the flow equations for compressible flow, this change in flow properties can be investigated.

11.2.2 *Equations for compressible flow*

If we assume that the density of a flow can vary, which it often does in reality, then the equations we developed in Chapter 2 need some modification. For the continuity of mass, the major change comes from allowing for the possibility of mass accumulating inside the patch of fluid due to the density changing with time. Also, the mass flow terms at each boundary of the patch must now include the density. This leads to the modified equation

$$\frac{\partial \rho}{\partial t} + \frac{\partial}{\partial x} (\rho u) + \frac{\partial}{\partial y} (\rho v) = 0 \tag{11.16}$$

and the momentum equations become (Liepmann and Roshko, 1957)

$$\frac{\partial}{\partial t} (\rho u) + \frac{\partial}{\partial x} (p + \rho u^2) + \frac{\partial}{\partial y} (\rho u v) = \frac{\partial \tau_{xx}}{\partial x} + \frac{\partial \tau_{xy}}{\partial y} \tag{11.17}$$

and

$$\frac{\partial}{\partial t} (\rho v) + \frac{\partial}{\partial x} (\rho u v) + \frac{\partial}{\partial y} (p + \rho v^2) = \frac{\partial \tau_{xy}}{\partial x} + \frac{\partial \tau_{yy}}{\partial y} \tag{11.18}$$

where the stress terms τ are known functions of the velocity gradients and the viscosity.

To model a compressible flow we must be able to describe the velocity field by its velocity components and we must also be able to specify the pressure and density. This means that we have to find four variables for the two-dimensional problem, and so the three equations above cannot give us enough information. To complete the mathematical definition of the problem, we can write an energy equation similar to equation (11.6) in the previous section which adds yet another variable, temperature.

As before, the two momentum equations (11.17) and (11.18) can be used to find the velocity components, and the compressible continuity equation (11.16) can then be used to obtain the fluid density throughout the flow. Temperature can be calculated using the energy equation, and finally the fluid pressure can be obtained from the equation of state for the fluid. Usually, the fluid can be taken as being a perfect gas, and so the equation of state is (see Rogers and

Mayhew, 1980)

$$p = \rho RT \tag{11.19}$$

where R is the gas constant.

From this, we can see that the compressible flow equations do not require a SIMPLE-like algorithm, as was discussed in Chapter 3. The equations derived allow the solution to proceed in a more intuitive way. Also, historically, people have solved the above equations with the viscous terms neglected. The equations are then known as the Euler equations. Once the local Mach number is greater than unity, the equations change in character to allow the features such as shock waves to occur. They become hyperbolic, and have characteristic solution directions. Numerical schemes capable of solving these equations must reflect these changes (Smith, 1985).

11.2.3 *Some practical problems with compressible flows*

Compressible flows, in reality, can exhibit behaviour that is very different from that of incompressible flows. This comes from the existence of shock waves in some flows, where the flow variables change rapidly over very small distances. Effectively, the flow solution is discontinuous. Also, if the flow speed is very large, when compared to the local speed of sound, the equations change in character to reflect the changes that occur in the physical flows.

This leads to CFD software having to be able to handle discontinuous flow solutions and different types of partial differential equation. Hopefully, the CFD software package will take care of the solution scheme changes, but the analyst must be aware of the requirements for a suitable mesh. This can be a problem, as the shock waves appear in the flow and propagate away from the boundaries, not along the boundaries. As the flow near a shock wave is discontinuous, the mesh must be built such that it has a large number of cells in areas where shock waves are expected. Unfortunately, before the solution is run, it is very difficult to predict where the shock wave will occur, and so one useful technique is to use adaptive meshing as described in Chapter 6. Then a solution can be run, the gradients of, say, the pressure calculated and the domain remeshed to put more cells in the areas of high gradients (see Zienkiewicz and Taylor, 1989).

11.3 Multiple-phase flows

Multiple-phase flows occur when two or more different states of material flow together. A solid might flow together with a gas or a liquid, or a gas might flow with a liquid. Physical examples of this are:

- the flow of steam and water in power plants;
- the flow of water droplets and air in a cooling tower;
- the flow of sand and air in a sand-transport system.

Modelling of these systems again uses the concepts of momentum conservation, continuity and, if necessary, energy conservation or other physical laws. In particular, each phase of material is assumed to have its own velocity components and a volume fraction. This latter quantity is the amount of material of a phase, by volume, relative to the total amount of material.

Equations are then developed for the conservation laws, and the interaction of the phases is taken into account by terms which are derived empirically. Take the case of solid spheres flowing with a liquid. The spheres gain momentum due to the relative velocity of the spheres when compared to the fluid, and the fluid loses momentum in the same way. Effectively, there is a friction between the phases, and this is known as the *interphase friction*. Consequently, the momentum equations have to be modified to allow for the effects of these interface momentum changes.

Standard papers on the modelling of multi-phase flows are those of Harlow and Amsden (1975) and Spalding (1980).

11.4 Modelling the effects of combustion

Combustion is the science of burning substances. It is a composite science, drawing on material from chemistry, thermodynamics and fluid mechanics. Much of the material that has already been discussed is of use in solving combustion problems, but there is a tremendous amount missing, especially in terms of chemistry. The following brief note might be of help if you need to solve flow problems which involve combustion.

There are several books that give good introductions to the subject, such as Spalding (1955). In terms of modelling, the simulation of combustion includes a combination of compressible flow and multi-phase flow, together with some chemistry which models the burning process. This takes some fuel, together with an oxidant, and produces what are known as the products of combustion. The rate at which this takes place has to be determined and is often controlled by the mixing of the components or by chemical kinetics, which are energy processes. The burning process releases heat energy to the system, and so the rate of heat release has also to be found. These features are calculated using simple models or empirical data.

12 Acquiring CFD technology

12.1 Preliminaries

In the first eleven chapters of this book, we have discussed the techniques that are used to produce a simulation of a flow using CFD. Also, we have looked at the hardware and software that is commercially available for use in the analysis process. For people who need to understand what is happening when a fluid flows, these techniques might provide an additional means of finding information, but the acquisition of the necessary hardware, software and expertise is expensive. For this reason, a careful study of the needs and requirements has to be made before the final decision to acquire the technology can be taken.

In this final chapter we look at some of the items that should be considered before committing resources to CFD. This chapter has been written to help guide those who have to make commercial decisions about the use of this technology, and so it is aimed mainly at the industrial user, but it also has some relevance for other users.

12.2 Assessing the need

The first thing to consider is how a knowledge of fluid flow might help you or your organization. To explore this, we need to look at all the areas that are related to fluid flow. Let us consider the case of a manufacturer in the motor industry. Fluid flow is an important topic for many different people within the organization, such as the vehicle aerodynamicist, the engine designer and the engineer developing heating and air conditioning systems. All of these areas are related to the production of the company product, but there are other areas where fluid flow could be important. For example, in the design of a new manufacturing facility, the flow of air could affect the manufacturing process, the ventilation or even the safety of the plant with regard to a fire.

Once the areas of interest have been listed, the techniques that are available to investigate the fluid flow phenomena should be assessed. In the case of the flow over a vehicle, it is cheaper and more accurate to obtain estimates of the aerodynamic forces that engineers require using experiments in a wind tunnel. Conversely, in the case of the flow within an engine, the experiments needed to determine the characteristics of a flow are extremely expensive due to the problems of measuring the flow variables, and so CFD might provide an alternative analysis tool that is cost effective. When all this information has been assembled, it should be possible to see if there is a place for CFD in the toolkit of techniques that might be used.

If CFD has a place in the toolkit, then the benefits of using this expensive technology must be made clear. Remember, it may be that CFD is just another tool, providing no more or no less than those tools currently in use. Conversely, it may provide some extra benefit, such as a direct saving in cash terms, or extra information that is not currently available.

12.3 Producing a specification for a CFD program

If there is a need for CFD, then we must decide what type of CFD software is required. To find out which of the available packages could be used, we must produce a list of requirements that the CFD software should meet. More often than not, no single package will meet all the requirements, but several packages will meet some of the requirements. Hence, when choosing the software we might have to make some very subjective decisions.

To draw up the program specification, we must think carefully about the flow problems that we wish to analyse. For example, we should have some idea of the following features:

- *The geometry of the fluid domains that might need to be analysed.* This will tell us whether we need a package that can solve problems in one, two or three dimensions. Also, if the geometry is very complex, we might be forced to use a system that can handle a mesh which is unstructured, but for many problems a structured mesh will be sufficient.
- *The flow type.* The classification of a flow depends on such things as the speed of the flow relative to the speed of sound,

so that an assessment of the compressibility of the fluid can be made. If the Mach number throughout the flow is low, then an incompressible flow solver can be used, but if the Mach number is close to or greater than one, anywhere in the flow, than a compressible solver will be required. A knowledge of the Reynolds number is also important, as we can determine from this whether the flow will be laminar or turbulent. If the flow is turbulent, then some form of turbulence model will be necessary, and the level of sophistication required of this model will also be determined by the characteristics of the flow. In most cases, a two-equation model such as the $k-\varepsilon$ model will be sufficient, but if the flow swirls, for example, then an algebraic stress or Reynolds stress model might well perform better. One other aspect of the flow that needs to be decided upon is the level of variation with time. Many flows can be assumed not to vary with time, provided that the gross features do not change with time, even though the microscopic flow features may vary with time. However, some flows will vary with time. This might be inherent in the flow itself even if the geometrical boundaries do not move. For example, the vortices can be shed behind a cylinder in a periodic manner at certain Reynolds numbers. In other cases, the flow will vary with time due to the movement of the geometrical boundaries. An example of this is the flow generated by a piston moving within an internal combustion engine.

- *Heat transfer effects.* In many flow situations, a knowledge of the flow of heat throughout the fluid might be required. Further, the flow of heat in adjoining solid material might also be of interest, requiring a conjugate heat transfer problem to be solved, as might the effects of radiation.
- *The number of phases in the flow.* This is usually one, but it could be two or more for some problems.

The above list covers some aspects of the flow, but we also need to determine some of the features of the simulations. For example, we should have some ideas about the following:

- *The size of the simulation problem.* We need to know something about the number of cells or elements that a typical mesh will contain and the number of flow variables that we need to calculate. This information helps us to determine the

219

storage requirements of the CFD programs in terms of both primary and secondary storage. It is worth remembering that the more data we calculate for a given flow, the more accurate the solution should be, but the longer it will take to obtain the results. Clearly some compromise has to be made here.

- *The required results of the analysis*, such as velocities, pressures or forces.
- *Interfacing requirements*. When defining the geometry of a flow domain, geometrical data is required and, sometimes, this will come from a CAD system or from a finite element pre-processor. Equally, we may wish to send the results to an existing post-processor or some other display software. If this is the case, then the CFD software should have appropriate interfaces.
- *Solution speed*. Many things affect the time that it takes to produce the solution to a simulation problem. Clearly, this will depend on the processing speed of the hardware that is used, but it also depends on the CFD solver itself. Some algorithms for solving the governing equations are much faster than others. This speed difference might come from the basic discretization of the equations, from the internal organization of the program or from the speed of the linear equation solvers used.
- *Hardware availability*. If there is a restriction on the make or type of computer or graphics terminal that you wish to run the software on, this should be noted. It is common for CFD pre- and post-processors to be very hardware-specific and for solvers to be much more portable between different machine types.

From all of the above, we now know a considerable amount about the flows that we wish to simulate and the simulation process itself. Finally, it is important to assess what kind of service it is that we need the software supplier to provide. This will be a very subjective set of requirements, and will to some extent depend on the people that are available within an organization to run the CFD software and talk with the supplier. At this point, it is worth issuing a warning. With the proliferation of computers and software, many people are now used to buying a package, loading it into a machine and getting results without too many problems. For business software this is certainly true, and it is becoming true of many engineering packages as well. Unfortunately, CFD software is not as mature as other engineering analysis tools that are on the market. Structural finite element packages that solve linear statics problems, and matrix manipulation packages, are much less prone to error than the latest CFD packages. This is because CFD technology is still developing and even the researchers in the area are not entirely sure as to how things will develop in the future. Further, it is only since, say, 1985 that industrial companies have started to take an interest in running CFD tools. This means that the wishes and demands of users have not yet been met in full.

Amongst the requirements that are related to the software supplier are the following:

- *Quality assurance or QA.* This is the extent to which the software has been tested against standard test cases for which there is a known solution. The comparison data may come from either analytical expressions or experiments. These comparisons are carried out both to verify the code, which means to show that it is correct against the numerical models that it is simulating, and also to validate the code, which is the process of showing that the code gives reliable results against physical experiments. There should be some evidence from the supplier that the software has been tested for both validation and verification. In fact, most pieces of CFD software are so complex that every possible combination of operating features will never be tested, until, that is, you as a user run your particular example, and find that it does not work. This may appear cynical but it is often true. Hopefully, as more people use CFD so the problems for users will reduce.
- *User friendliness.* This is probably the most subjective feature of all, as what appears friendly to one person will be unfriendly to another. Again, this will depend on the staff who run the programs.
- *User support.* This is very important as users can never be fully conversant with the programs that they run. Software suppliers should provide some form of User Hotline that can give a quick response to a user's questions. This normally comes as part of the annual licence fee for the software, or it can be purchased separately if the program is bought with

Table 12.1. *Product specification*

Capability	Prog. 1	Prog. 2	Prog. 3	...	Prog. N
Dimensions (1,2,3)	...	...	...	...	...
Mesh type (structured or unstructured)	...	...	...	...	...
Turbulence model	...	...	...	...	...
Compressible (yes or no)	...	...	...	...	...
Steady or unsteady flow	...	...	...	...	...
Moving geometry (yes or no)	...	...	...	...	...
Heat transfer in fluid (yes or no)	...	...	...	...	...
Heat transfer in solid (yes or no)	...	...	...	...	...
Fluid phases (1 or 2 or more)	...	...	...	...	...
Limits on model size	...	...	...	...	...
Interfacing to CAD	...	...	...	...	...
Speed of solution	...	...	...	...	...
Hardware availability	...	...	...	...	...
Evidence of QA	...	...	...	...	...
User friendliness	...	...	...	...	...
User support	...	...	...	...	...
List of users with similar flow problems	...	...	...	...	...

a once-only payment, which is known as a 'perpetual licence'. There should also be the option of buying training in the use of the programs, and the chance for users to work with the supplier in setting up a problem. This is normally done by paying for consultancy from the software supplier.

- *Current users*. It is important to know who is currently using the software, not who has used the software. This enables companies to see if firms in a similar business are using the software, and can give some confidence in the suppliers and their products.

The easiest way to document this information is to draw up a table of capabilities product by product. Sometimes this can be difficult to do for someone with little experience. It is at this stage that it is important to obtain independent advice to guide you. Sometimes people place too much reliance in the software suppliers themselves, and even though the suppliers can provide much information, an independent view is worth while. A sample specification table is shown in Table 12.1 and this can be used to assess each of the competing products.

12.4 Deciding on the necessary software

Once the specification of the software is determined, various software options need to be evaluated against this specification. This can take a lot of time and effort as there are many CFD products in the marketplace and the suppliers of each of them will be only too willing to shower you with information. The information that is provided can take many forms, but the simplest starting point is to look at the brochures that explain the software. Much of the information required can be determined from these, but quite a lot of it cannot. In particular, the more subjective information, such as the levels of user friendliness, solution times, QA and user support, needs to be investigated further.

One way of gaining this more specific information is to produce a sample problem that is typical of the problems you wish to solve. Suppliers will often produce a simulation of this problem using their software at a reduced cost, or even for free if the problem is very small. This enables potential customers to see software products in action on a realistic problem. Such a trial will help in understanding how the processes outlined in this book relate to the specification and operation of the software. It will also produce some hard facts that should help in determining the cost of obtaining a simulation using a particular CFD package.

When the competing products have been assessed using the specification table, several suitable products should emerge. In fact it is probable that none of the products will be ideal, but some should come closer than others. A simple way of assessing the most suitable package is to assign numbers to each of the categories in the specification in some way such that the higher the number the better the specification level. Then they can be ranked by adding up these numbers and getting a total value for each package.

Once the products are ranked in order of suitability, the question of cost needs to be looked at. Normally, software is licensed on an annual basis with a single fee being paid to the supplier which includes the provision of the software and any updates to it, as well as technical support in the form of a hotline service. Sometimes, however, the software is purchased on perpetual licence terms where one large payment pays for the software and a smaller annual fee pays for the updates to the software and the support. Sometimes, both methods are on offer, and it takes careful consideration to decide which of the two will be the cheapest option in the long run.

This is especially difficult as the market is still developing and the most suitable program today may not be the best choice in three or four years' time. Finally, it may be that some sacrifice in terms of the capability of a package has to be made if an affordable solution is to be chosen for purchase. This has to be achieved by determining the minimum level of functionality that is acceptable.

12.5 Deciding on the necessary hardware

Many organizations already have access to the computer facilities that are necessary for running large computational analysis programs such as CFD packages. Others will need to acquire the hardware. In both cases, however, it is important to consider a number of factors. For the former case this enables the user to determine if the existing facilities are suitable and have the necessary spare capacity, and for the latter case it allows estimates to be made of the various measures that will determine the hardware.

These factors include the following:

- *Computer processing power.* A large amount of processing power is needed to run some CFD test cases. Fortunately, recent technical advances mean that the necessary power is available very cheaply. The factors that affect the speed of processing include such things as the calculation speed of the processor which is measured in *mips* and is the number of millions of processor instructions carried out per second, or in *MFLOPS* where one *MFLOP* is one million floating point operations per second. There is no clear relationship between the two for different processors, as what takes one instruction on one machine might take several instructions on another. The speeds of the various computers are often quoted in these units, but different software runs in different ways on different machines. Consequently, the numbers quoted are only a guide to the raw processing power. To find a true measure of speed for the software and hardware combination, a series of sample flow problems must be simulated. This assumes that the CFD software does not make any use of the secondary data storage during execution, as the speed at which data can be accessed from devices such as hard disks can have a marked effect on solution times. Some CFD software packages write data to these devices during the solution phase, and if the processes of reading and writing to the disk are slow, then the whole solution process is slowed down. For a typical analysis on a given computer installation, the total solution time will depend on all of these things together with the number of simulations that will be solved simultaneously on any one system.

- *Primary data storage capacity.* On most computer systems, the primary data storage system is known as 'random access memory' (RAM). This is usually sized by the number of bytes of data that can be stored. Each byte consists of eight bits, where one bit is the basic unit of storage corresponding to a stored value of either zero or one. Numbers can be stored as integers or real numbers; two or four bytes are used for integers and four or eight bytes for real numbers. The greater the number of bytes the greater the maximum integer, and the more accurate a real number that can be stored. Sometimes, the software supplier will specify the number of megabytes of RAM that are required to run their software successfully. In large machines, such as supercomputers, the memory size is measured in words. These are usually words of eight bytes or sixty-four bits, and are the machine's minimum storage for a single real number.

- *Secondary storage capacity.* Random access memory is used during the execution of a program, but if the user needs to access the data after the program has stopped running, then the data must be written to some secondary storage device. These are usually hard disks, which are aluminium disks covered in magnetic material such as iron oxide, as in audio tape. In personal computers these disks may store a few tens of megabytes and in workstations several hundred megabytes. In large systems, the disk storage might consist of sets of disks, each storing several gigabytes of data. We need to assess how much of this storage we will need for each problem that we wish to solve. A rough estimate can be made by taking the number of nodes in a problem and multiplying by the number of coordinates used to describe a node plus the number of variables stored at each point, which would be nine for a three-dimensional turbulent flow problem solved with three velocity components, pressure and two turbulence variables. So for a mesh with 10 000 nodes we must store at least 90 000 real numbers. If the data is stored in readable

(ASCII) format, say twenty bytes are required to store each number, 1.8 megabytes are required in total. If, however, the data is stored as single precision real numbers in binary format, only four bytes will be required to store each number and the total storage required will be 0.36 megabytes. These are low estimates of the total data storage requirements, as each software package will store different information. The software supplier might be able to give information on the data storage required for a given model size.

- *Access points*. If several people need to run CFD analyses simultaneously then several access points will be required. These might need to be split between a number of graphics screens and a number of text screens. This will enable some people to perform graphics pre- and post-processing whilst others run the solver program.

- *Backup facilities*. There is a need to provide some backup of the data held on disk, to protect against loss of data. This can occur if a disk drive is broken, such that the data stored on it cannot be read, or could occur if a user deletes a file in error. It is common for each disk to be backed-up in full, i.e. all the data is written to a tape storage device, or something similar, every week. Then further backup procedures are carried out once a day to ensure that all the new files that have been created within the previous twenty-four hours, and the new versions of edited files, are also written to a backup device. This procedure is known as an 'incremental backup' and ensures that, at worst, only one day's work can be lost. Once backup tapes have been prepared it is worth protecting them against fire by using a fireproof storage facility.

When these items have been considered, it should be possible to know whether an existing installation will be sufficient to run CFD problems or whether it will need to be enhanced in some way. If new facilities are required, either to enhance the existing capacity or to provide a completely new system, then they could now be assessed for suitability.

12.6 Finding people to run CFD simulations

Having decided upon a software package and a hardware system, CFD simulations will not run themselves. We must, finally, look at

226

the most important asset in the CFD analysis process. This is the analyst who actually translates the engineering problem into a computational simulation, runs the CFD solver and analyses the results. It is the skill of this person, or set of persons, that will determine whether all the hardware and software will be utilized in the best possible way and produce good quality results.

People from many different backgrounds can be trained to go through the processes that we have discussed in this book, but it is my personal belief that more than this is required. The skills that are required include the following:

- *Mathematical skills*. These enable the analyst to understand the underlying features of the numerical processes used to convert the governing partial differential equations into numerical analogues, and to coax the solution procedures to converge to sensible and realistic values.
- *Computational skills*. The production of a CFD simulation can involve the user in manipulating large amounts of data with packages that do not interface together and that reside on a variety of types of computers. This can mean, for example, that CFD analysts have to write their own interface programs to convert data from one program's format to another program's format. Also, an analyst might have to write computer operating system command language programs that instruct a computer, or even a variety of computers, to move data around a network, run some CFD programs and then move the data around the network again. Consequently, CFD analysts must be conversant with computer procedures at a level that is far greater than that required for analysts who use the more common software products that perform engineering computations.
- *Good interpersonal skills*. If the analyst is not the end-user of the data, then there will have to be close liaison between the analyst and the end-user, who is in effect a customer or client of the analyst. This requires that a good working relationship is developed between the two parties so that the analyst knows what the customer requires, and the customer is aware of the limitations of the analysis.
- *Engineering skills*. Finally, the analyst must have a working understanding of the engineering processes that are to be modelled. This enables the limits of a computer model to be

established and the results of the simulation to be analysed in a sensible way.

Large organizations may well have a pool of analysts in which there are several people that could be used to produce CFD simulations, as they have a majority of the qualities listed above. These people could be engaged at present in running finite element structural analyses or similar large scale engineering computations. If such people do not exist within the organization, or if suitable people cannot be used for whatever reason, then staff will have to be hired. Hiring staff of the right technical background to use CFD in industry, whatever their background, is extremely difficult. Not many people have all the necessary skills and so several people may be needed. Depending on the size of the organization, therefore, one or more people may be employed in the use of CFD, and the right mixture of abilities is important.

One other way of proceeding is to employ a limited number of people to work with CFD and then to use external consultants to supplement the skills where appropriate. These consultants can be found working with CFD software suppliers, general engineering consultancy practices and in universities and polytechnics. For industrial users who are not specialists in this field, it is important to have access to advice at a moment's notice. This can be provided by a software supplier when problems occur in running a particular package, but another useful source is a local university or polytechnic, where a specialist in the CFD field may well be willing to provide consultancy as and when required.

12.7 Integrating CFD within the design process

As a final topic, let us look at how the results of CFD simulations can be used within the engineering design process. In industry, CFD can be used to provide information about how fluids flow and what the effect of the flow is on engineering devices. Chapter 1 gave a list of possible uses. At present, there are many situations in which CFD simulations are being made, and many examples have been presented at technical conferences, or in journals, which demonstrate the benefits of using CFD technology in particular industrial areas.

Now, whilst these demonstrations are useful in that they show a technical capability, their impact on the engineering design process is limited. If CFD is to be of use to an industrial organization, then

228

it should be capable of being integrated into the design process in such a way that the simulations can influence the engineering design. This could happen after a given design has been proposed and before prototypes are built. However, this can only take place if the CFD simulation can provide the required engineering data in a cost-effective way when compared to current methods of analysis, whatever they are, and in a shorter span of time.

At present, the total integration of CFD into the design process is not possible because it takes a long time, perhaps several man-months, to build a computer model of a flow problem, run the solver and produce the results. All of this must be repeated for every configuration that is considered. Compare this to the use of a physical model. Whilst it may take some time to build the model, once built it can be tested at a variety of flow conditions and for a variety of geometrical configurations with a minimum of extra effort.

Looking at the CFD process that we have discussed, the generation of the mesh takes the longest time for complex but realistic flow geometries. If this mesh building time can be reduced, which it can be if automatic mesh generation tools are developed further, then the turnaround time for a CFD simulation can be reduced to a few hours. Once the results of CFD simulations can be accessed in such a short space of time for each configuration, then the design process can be influenced much more easily as several configurations can be carried out and modifications made to the design, which can be quickly modelled by the CFD analyst and tested computationally. When this is possible, then CFD will be a mature technology for use in industry. At present (1991), CFD is a useful tool but there is still room for improvement.

Appendix: PHOENICS results for a simple laminar flow

Results After 100 Sweeps

```
****************************************************************


   ------------------------------------------------------------
        CCCC HHH         PHOENICS - EARTH      Version 1.5.3
     CCCCCCCC HHHHH      (C) Copyright 1989
    CCCCCCC HHHHHHHHHH   Concentration Heat and Momentum Ltd
   CCCCCCC HHHHHHHHHHHH  All rights reserved.
   CCCCCC HHHHHHHHHHHHHH CHAM Ltd, Bakery House, 40 High St
   CCCCCCC HHHHHHHHHHHH  Wimbledon, London, SW19 5AU
    CCCCCCC HHHHHHHHHH   Tel: 01-947-7651; Telex: 928517
     CCCCCCCC HHHHH      Facsimile: 01-879-3497
        CCCC HHH         The option level is   -18

...edited...

****************************************************************
 Group 1. Run Title and Number
****************************************************************
****************************************************************


TEXT(SIMPLE DEVELOPING FLOW IN BETWEEN PLATES)

****************************************************************
****************************************************************


IRUNN   =        1 ;LIBREF =        0

*** GRID-GEOMETRY INFORMATION ***
X-COORDINATES OF THE CELL CENTRES
   5.000E-01
Y-COORDINATES OF THE CELL CENTRES
   2.500E-02  7.500E-02  1.250E-01  1.750E-01  2.250E-01
   2.750E-01  3.250E-01  3.750E-01  4.250E-01  4.750E-01
Z-COORDINATES OF THE CELL CENTRES
   1.953E-02  5.860E-02  1.172E-01  2.344E-01  4.688E-01
   9.375E-01  1.875E+00  3.750E+00  7.500E+00  1.500E+01

--- INTEGRATION OF EQUATIONS BEGINS ---
```

```
*************************************************************
TIME STP=    1 SWEEP NO=   100 ZSLAB NO=    2 ITERN NO=     1

*************************************************************
TIME STP=    1 SWEEP NO=   100 ZSLAB NO=    1 ITERN NO=     1

FLOW FIELD AT ITHYD=    1, ISWEEP= 100, ISTEP=   1
YZPR  IX=       1
FIELD VALUES OF P1
IY=  10  1.388E+02  1.236E+02  1.622E+02  1.595E+02  1.553E+02
IY=   9  1.371E+02  1.235E+02  1.606E+02  1.586E+02  1.551E+02
IY=   8  1.376E+02  1.248E+02  1.602E+02  1.584E+02  1.551E+02
IY=   7  1.385E+02  1.261E+02  1.601E+02  1.585E+02  1.550E+02
IY=   6  1.395E+02  1.271E+02  1.603E+02  1.587E+02  1.550E+02
IY=   5  1.405E+02  1.280E+02  1.609E+02  1.589E+02  1.550E+02
IY=   4  1.412E+02  1.287E+02  1.620E+02  1.593E+02  1.550E+02
IY=   3  1.421E+02  1.281E+02  1.638E+02  1.595E+02  1.548E+02
IY=   2  1.483E+02  1.210E+02  1.668E+02  1.592E+02  1.547E+02
IY=   1  1.662E+02  9.704E+01  1.702E+02  1.582E+02  1.545E+02
IZ=       1          2          3          4          5
IY=  10  1.490E+02  1.373E+02  1.144E+02  7.173E+01  4.170E-11
IY=   9  1.490E+02  1.373E+02  1.144E+02  7.174E+01  6.545E-11
IY=   8  1.490E+02  1.373E+02  1.144E+02  7.174E+01  6.328E-11
IY=   7  1.490E+02  1.373E+02  1.144E+02  7.174E+01  6.004E-11
IY=   6  1.490E+02  1.373E+02  1.144E+02  7.174E+01  5.578E-11
IY=   5  1.490E+02  1.373E+02  1.144E+02  7.174E+01  5.056E-11
IY=   4  1.490E+02  1.373E+02  1.143E+02  7.173E+01  4.451E-11
IY=   3  1.490E+02  1.373E+02  1.143E+02  7.173E+01  3.778E-11
IY=   2  1.490E+02  1.373E+02  1.143E+02  7.173E+01  3.062E-11
IY=   1  1.490E+02  1.373E+02  1.143E+02  7.173E+01  5.032E-11
IZ=       6          7          8          9         10
FIELD VALUES OF V1
IY=   9  1.140E-01  1.174E-01  7.076E-02  4.693E-02  1.220E-02
IY=   8  2.043E-01  1.978E-01  1.427E-01  9.490E-02  2.391E-02
IY=   7  2.719E-01  2.615E-01  2.092E-01  1.403E-01  3.344E-02
IY=   6  3.179E-01  3.072E-01  2.657E-01  1.786E-01  3.925E-02
IY=   5  3.412E-01  3.324E-01  3.094E-01  2.048E-01  4.005E-02
IY=   4  3.391E-01  3.317E-01  3.380E-01  2.133E-01  3.512E-02
IY=   3  3.157E-01  2.927E-01  3.475E-01  1.982E-01  2.482E-02
IY=   2  3.067E-01  1.773E-01  3.294E-01  1.541E-01  1.123E-02
IY=   1  3.592E-01 -8.412E-02  2.657E-01  8.187E-02 -8.201E-04
IZ=       1          2          3          4          5
IY=   9  5.606E-04 -5.055E-04 -1.038E-03 -7.173E-04 -2.708E-03
IY=   8  1.103E-03 -9.909E-04 -2.092E-03 -1.783E-03 -2.709E-03
IY=   7  1.469E-03 -1.441E-03 -3.060E-03 -2.712E-03 -2.710E-03
IY=   6  1.546E-03 -1.835E-03 -3.885E-03 -3.441E-03 -2.711E-03
IY=   5  1.287E-03 -2.147E-03 -4.499E-03 -3.904E-03 -2.712E-03
IY=   4  7.267E-04 -2.342E-03 -4.823E-03 -4.028E-03 -2.712E-03
IY=   3  2.324E-06 -2.371E-03 -4.751E-03 -3.735E-03 -2.712E-03
IY=   2 -6.434E-04 -2.157E-03 -4.130E-03 -2.948E-03 -2.712E-03
IY=   1 -8.519E-04 -1.533E-03 -2.689E-03 -1.636E-03 -2.712E-03
IZ=       6          7          8          9         10
FIELD VALUES OF W1
IY=  10   1.089E+00  1.162E+00  1.282E+00  1.428E+00  1.505E+00
IY=   9   1.070E+00  1.124E+00  1.249E+00  1.400E+00  1.474E+00
IY=   8   1.053E+00  1.097E+00  1.212E+00  1.353E+00  1.414E+00
IY=   7   1.036E+00  1.070E+00  1.169E+00  1.287E+00  1.324E+00
IY=   6   1.018E+00  1.039E+00  1.118E+00  1.198E+00  1.204E+00
IY=   5   9.984E-01  1.004E+00  1.057E+00  1.083E+00  1.052E+00
IY=   4   9.817E-01  9.604E-01  9.809E-01  9.321E-01  8.672E-01
IY=   3   9.929E-01  9.059E-01  8.749E-01  7.344E-01  6.488E-01
IY=   2   1.041E+00  8.186E-01  7.001E-01  4.720E-01  3.961E-01
IY=   1   7.194E-01  8.144E-01  3.617E-01  1.120E-01  1.174E-01
IZ=       1          2          3          4          5
```

Appendix

```
IY=  10     1.512E+00  1.499E+00  1.447E+00  1.376E+00
IY=   9     1.481E+00  1.469E+00  1.416E+00  1.309E+00
IY=   8     1.418E+00  1.407E+00  1.359E+00  1.266E+00
IY=   7     1.325E+00  1.315E+00  1.274E+00  1.201E+00
IY=   6     1.200E+00  1.193E+00  1.162E+00  1.116E+00
IY=   5     1.045E+00  1.040E+00  1.024E+00  1.011E+00
IY=   4     8.581E-01  8.574E-01  8.610E-01  8.903E-01
IY=   3     6.406E-01  6.459E-01  6.770E-01  7.557E-01
IY=   2     3.934E-01  4.090E-01  4.810E-01  6.123E-01
IY=   1     1.278E-01  1.659E-01  3.003E-01  4.640E-01
IZ=         6          7          8          9

...edited...

************************************************************
SPOT VALUES VS. SWEEP (/ITHYD IF PARAB)
   IXMON=        1    IYMON=        2    IZMON=        2

TABULATION OF ABSCISSA AND ORDINATES...
      ISWP        P1         V1         W1
  1.000E+00  1.000E-10  4.053E-02  9.950E-01
  2.000E+00  1.391E-01  1.067E-01  9.617E-01
  3.000E+00  3.632E-01  1.313E-01  9.370E-01
  4.000E+00  6.851E-01  1.696E-01  9.186E-01
  5.000E+00  1.106E+00  1.816E-01  9.050E-01
  6.000E+00  1.623E+00  2.056E-01  8.906E-01
  7.000E+00  2.239E+00  2.109E-01  8.793E-01
  8.000E+00  2.951E+00  2.274E-01  8.670E-01
  9.000E+00  3.736E+00  2.315E-01  8.590E-01
  1.000E+01  4.601E+00  2.419E-01  8.490E-01

...edited...

  9.100E+01  1.123E+02  1.790E-01  8.228E-01
  9.200E+01  1.133E+02  1.786E-01  8.224E-01
  9.300E+01  1.143E+02  1.782E-01  8.221E-01
  9.400E+01  1.153E+02  1.779E-01  8.217E-01
  9.500E+01  1.162E+02  1.777E-01  8.212E-01
  9.600E+01  1.172E+02  1.775E-01  8.207E-01
  9.700E+01  1.182E+02  1.773E-01  8.203E-01
  9.800E+01  1.191E+02  1.773E-01  8.197E-01
  9.900E+01  1.201E+02  1.772E-01  8.192E-01
  1.000E+02  1.210E+02  1.773E-01  8.186E-01
  VARIABLE        P1         V1         W1
    MINVAL=  1.000E-10  4.053E-02  7.920E-01
    MAXVAL=  1.210E+02  2.747E-01  9.950E-01
    CELLAV=  5.988E+01  2.208E-01  8.211E-01
```

PHOENICS results for a simple laminar flow

```
1.00  W....+...VVVVVVVV...+....+....+....+....+....+..PPP
      .        VV       VVVV                      PPPP .
0.90  +     VV          VVV                   PPPP     +
      .W   V            VVV                   PPP      .
0.80  +    V          VVV        PPP                   +
      .   V          VVV     PPP                       .
0.70  +W V                  VVVV                       +
      .                PPP  VVVVV                      .
0.60  + W            PPP          VVVVVVVVVVV           +
      . W          PPP                                 .
0.50  + W                  PP                          +
      . W                 PP                           .
0.40  +V                 PP                            +
      .  W              PP                             .
0.30  +V   W          PP                               +
      .    W        PP                                 .
0.20  +    W      PP                                   +
      .     W   PP          WWWWWWWWWWWWWWWWWWWWW       .
0.10  +     PWW       WWWWWW                           +
      .   PPPP WWW      WWWWWW                          .
0.00  VPPPP+....+WWWWWWWWWW....+....+....+....+....+....+
      0    .1   .2   .3   .4   .5   .6   .7   .8   .9  1.0
THE ABSCISSA IS      ISWP.  MIN= N1.00E+00 MAX= 1.00E+02
********************************************************
```

```
********************************************************
RESIDUALS VS. SWEEP (/ITHYD IF PARAB)

TABULATION OF ABSCISSA AND ORDINATES...
     ISWP        P1         V1         W1
   1.000E+00  2.756E+06  6.753E+04  8.983E+08
   2.000E+00  2.733E+06  2.124E+06  1.016E+09
   3.000E+00  2.465E+06  2.567E+06  1.013E+09
   4.000E+00  2.779E+06  3.106E+06  1.509E+09
   5.000E+00  2.895E+06  3.011E+06  1.463E+09
   6.000E+00  2.636E+06  3.871E+06  2.384E+09
   7.000E+00  2.669E+06  4.324E+06  1.043E+09
   8.000E+00  2.804E+06  3.729E+06  1.298E+09
   9.000E+00  2.676E+06  3.791E+06  1.770E+09
   1.000E+01  2.664E+06  3.973E+06  9.224E+08

...edited...

   9.000E+01  1.052E+06  2.123E+06  3.499E+08
   9.100E+01  1.037E+06  2.095E+06  3.463E+08
   9.200E+01  1.022E+06  2.066E+06  3.422E+08
   9.300E+01  1.007E+06  2.038E+06  3.387E+08
   9.400E+01  9.934E+05  2.010E+06  3.350E+08
   9.500E+01  9.790E+05  1.981E+06  3.311E+08
   9.600E+01  9.646E+05  1.953E+06  3.274E+08
   9.700E+01  9.512E+05  1.926E+06  3.240E+08
   9.800E+01  9.372E+05  1.898E+06  3.202E+08
   9.900E+01  9.239E+05  1.871E+06  3.170E+08
   1.000E+02  9.109E+05  1.845E+06  3.136E+08
   VARIABLE      P1         V1         W1
     MINVAL=  1.372E+01  1.112E+01  1.956E+01
     MAXVAL=  1.488E+01  1.532E+01  2.159E+01
```

233

```
1.00 +.PW.+VVVVVVVVVVVVVV.+....+....+....+....+....+....+
     PPPVVVPVP P       VVVVVVVVVV                      .
0.90 + VP  PP PPPP          VVVVVVVVV                  +
     .V  W      PPP              VVVVVVVVV              .
0.80 +V          PPP                     VVVVVV        +
     . W   W      PPP                                  .
0.70 +   W  W W      PPP                               +
     .         W      PPP                              .
0.60 +W W W W          PPP                             +
     .   W  W            PPP                           .
0.50 W       W W          PP                           +
     .     W   W           PP                          .
0.40 +    WW WWWWW           PP                         +
     .       WWWWW            PP                       .
0.30 +        WWWW             PP                       +
     .         WWWW              PP                    .
0.20 +          WWWWW             PP                    +
     .           WWWWW             PP                  .
0.10 +                    WWWWW PP                      +
     .                     WWWWWP .
0.00 V....+....+....+....+....+....+....+....+....+..WWW
     0   .1   .2   .3   .4   .5   .6   .7   .8   .9  1.0
THE ABSCISSA IS      ISWP.  MIN= 1.00E+00 MAX= 1.00E+02
*********************************************************

*********************************************************
SATLIT RUN NUMBER =   1 ; LIBRARY REF.=   0
RUN COMPLETED AT 16:49:16 ON TUESDAY, 27 NOVEMBER 1990
MACHINE-CLOCK TIME OF RUN =      82 SECONDS.
TIME/(VARIABLES*CELLS*TSTEPS*SWEEPS*ITS) = 2.733E-03
*********************************************************
```

Results After 500 Sweeps

```
*********************************************************

---------------------------------------------------------
      CCCC HHH         PHOENICS - EARTH     Version 1.5.3
   CCCCCCCC HHHHH      (C) Copyright 1989
  CCCCCCC HHHHHHHHHH   Concentration Heat and Momen-
                       tum Ltd
 CCCCCCC HHHHHHHHHHHH  All rights reserved.
 CCCCC HHHHHHHHHHHHHH  CHAM Ltd, Bakery House, 40 High St
 CCCCCCC HHHHHHHHHHHH  Wimbledon, London, SW19 5AU
  CCCCCCC HHHHHHHHHH   Tel: 01-947-7651; Telex: 928517
   CCCCCCCC HHHHH      Facsimile: 01-879-3497
      CCCC HHH         The option level is      -18
---------------------------------------------------------

...edited...

*********************************************************
 Group 1. Run Title and Number
*********************************************************
*********************************************************

TEXT(SIMPLE DEVELOPING FLOW IN BETWEEN PLATES)

*********************************************************
```

234

PHOENICS results for a simple laminar flow

```
*****************************************************************

IRUNN   =      1 ;LIBREF =      0

*** GRID-GEOMETRY INFORMATION ***
X-COORDINATES OF THE CELL CENTRES
   5.000E-01
Y-COORDINATES OF THE CELL CENTRES
   2.500E-02  7.500E-02  1.250E-01  1.750E-01  2.250E-01
   2.750E-01  3.250E-01  3.750E-01  4.250E-01  4.750E-01
Z-COORDINATES OF THE CELL CENTRES
   1.953E-02  5.860E-02  1.172E-01  2.344E-01  4.688E-01
   9.375E-01  1.875E+00  3.750E+00  7.500E+00  1.500E+01

--- INTEGRATION OF EQUATIONS BEGINS ---


*****************************************************************
TIME STP=     1 SWEEP NO=   400 ZSLAB NO=    2 ITERN NO=     1

*****************************************************************
TIME STP=     1 SWEEP NO=   400 ZSLAB NO=    1 ITERN NO=     1

FLOW FIELD AT ITHYD=   1, ISWEEP= 400, ISTEP=   1
YZPR  IX=      1
FIELD VALUES OF P1
IY= 10  1.858E+02  1.879E+02  1.890E+02  1.865E+02  1.821E+02
IY=  9  1.840E+02  1.860E+02  1.873E+02  1.855E+02  1.819E+02
IY=  8  1.846E+02  1.865E+02  1.875E+02  1.855E+02  1.819E+02
IY=  7  1.855E+02  1.872E+02  1.878E+02  1.855E+02  1.819E+02
IY=  6  1.867E+02  1.882E+02  1.882E+02  1.856E+02  1.819E+02
IY=  5  1.886E+02  1.896E+02  1.886E+02  1.856E+02  1.818E+02
IY=  4  1.913E+02  1.915E+02  1.890E+02  1.855E+02  1.818E+02
IY=  3  1.958E+02  1.940E+02  1.892E+02  1.853E+02  1.817E+02
IY=  2  2.036E+02  1.968E+02  1.885E+02  1.848E+02  1.815E+02
IY=  1  2.085E+02  1.894E+02  1.860E+02  1.835E+02  1.813E+02
IZ=     1          2          3          4          5
IY= 10  1.760E+02  1.642E+02  1.407E+02  9.360E+01  7.349E-11
IY=  9  1.760E+02  1.642E+02  1.407E+02  9.361E+01  7.396E-11
IY=  8  1.760E+02  1.642E+02  1.407E+02  9.361E+01  7.085E-11
IY=  7  1.760E+02  1.642E+02  1.407E+02  9.361E+01  6.617E-11
IY=  6  1.760E+02  1.642E+02  1.407E+02  9.361E+01  5.994E-11
IY=  5  1.760E+02  1.642E+02  1.407E+02  9.360E+01  5.216E-11
IY=  4  1.760E+02  1.642E+02  1.407E+02  9.360E+01  4.283E-11
IY=  3  1.760E+02  1.642E+02  1.407E+02  9.360E+01  3.197E-11
IY=  2  1.760E+02  1.642E+02  1.407E+02  9.360E+01  1.963E-11
IY=  1  1.760E+02  1.642E+02  1.407E+02  9.360E+01  8.367E-12
IZ=     6          7          8          9          10
FIELD VALUES OF V1
IY=  9  1.102E-01  1.067E-01  9.143E-02  4.990E-02  6.791E-03
IY=  8  2.154E-01  2.078E-01  1.777E-01  9.651E-02  1.296E-02
IY=  7  3.160E-01  3.035E-01  2.559E-01  1.349E-01  1.677E-02
IY=  6  4.117E-01  3.919E-01  3.217E-01  1.604E-01  1.695E-02
IY=  5  5.011E-01  4.694E-01  3.688E-01  1.690E-01  1.284E-02
IY=  4  5.799E-01  5.285E-01  3.880E-01  1.573E-01  4.698E-03
IY=  3  6.381E-01  5.532E-01  3.670E-01  1.247E-01 -6.082E-03
IY=  2  6.495E-01  5.102E-01  2.930E-01  7.681E-02 -1.681E-02
IY=  1  5.318E-01  3.364E-01  1.677E-01  3.063E-02 -2.361E-02
IZ=     1          2          3          4          5
IY=  9  6.101E-04  2.545E-05 -5.636E-06 -6.392E-05 -2.030E-04
IY=  8  1.169E-03  4.972E-05 -1.117E-05 -1.255E-04 -2.031E-04
IY=  7  1.575E-03  6.988E-05 -1.640E-05 -1.826E-04 -2.032E-04
IY=  6  1.763E-03  8.401E-05 -2.107E-05 -2.329E-04 -2.032E-04
```

235

```
IY=   5  1.718E-03   9.129E-05 -2.493E-05 -2.737E-04 -2.033E-04
IY=   4  1.489E-03   9.196E-05 -2.769E-05 -3.016E-04 -2.034E-04
IY=   3  1.175E-03   8.692E-05 -2.898E-05 -3.122E-04 -2.034E-04
IY=   2  8.866E-04   7.690E-05 -2.818E-05 -2.976E-04 -2.034E-04
IY=   1  6.646E-04   6.028E-05 -2.380E-05 -2.378E-04 -2.034E-04
IZ=       6           7          8          9          10
FIELD VALUES OF W1
IY=  10     1.086E+00  1.173E+00  1.310E+00  1.466E+00  1.509E+00
IY=   9     1.082E+00  1.164E+00  1.293E+00  1.439E+00  1.478E+00
IY=   8     1.079E+00  1.156E+00  1.273E+00  1.393E+00  1.417E+00
IY=   7     1.075E+00  1.147E+00  1.245E+00  1.324E+00  1.326E+00
IY=   6     1.070E+00  1.133E+00  1.202E+00  1.229E+00  1.204E+00
IY=   5     1.062E+00  1.110E+00  1.136E+00  1.099E+00  1.049E+00
IY=   4     1.045E+00  1.067E+00  1.030E+00  9.287E-01  8.618E-01
IY=   3     1.009E+00  9.768E-01  8.577E-01  7.084E-01  6.417E-01
IY=   2     9.081E-01  7.732E-01  5.760E-01  4.322E-01  3.896E-01
IY=   1     5.845E-01  3.088E-01  6.953E-02 -2.636E-02  1.114E-01
IZ=         1          2          3          4          5
IY=  10     1.516E+00  1.517E+00  1.517E+00  1.510E+00
IY=   9     1.485E+00  1.486E+00  1.485E+00  1.479E+00
IY=   8     1.422E+00  1.423E+00  1.423E+00  1.417E+00
IY=   7     1.328E+00  1.329E+00  1.329E+00  1.323E+00
IY=   6     1.203E+00  1.203E+00  1.203E+00  1.199E+00
IY=   5     1.046E+00  1.046E+00  1.046E+00  1.043E+00
IY=   4     8.579E-01  8.577E-01  8.577E-01  8.566E-01
IY=   3     6.381E-01  6.379E-01  6.379E-01  6.394E-01
IY=   2     3.869E-01  3.864E-01  3.866E-01  3.926E-01
IY=   1     1.032E-01  1.017E-01  1.029E-01  1.266E-01
IZ=         6          7          8          9
...edited...

*************************************************************
SPOT VALUES VS. SWEEP (/ITHYD IF PARAB)
  IXMON=      1   IYMON=      2   IZMON=      2

TABULATION OF ABSCISSA AND ORDINATES...
     ISWP       P1         V1         W1
  2.000E+00  1.219E+02  1.789E-01  8.181E-01
  3.000E+00  1.228E+02  1.756E-01  8.178E-01
  4.000E+00  1.237E+02  1.777E-01  8.170E-01
  5.000E+00  1.246E+02  1.754E-01  8.170E-01
  6.000E+00  1.254E+02  1.774E-01  8.159E-01
  7.000E+00  1.263E+02  1.760E-01  8.157E-01
  8.000E+00  1.271E+02  1.777E-01  8.146E-01
  9.000E+00  1.280E+02  1.770E-01  8.142E-01
  1.000E+01  1.288E+02  1.784E-01  8.131E-01
...edited...

  3.900E+02  1.968E+02  5.108E-01  7.717E-01
  3.910E+02  1.968E+02  5.108E-01  7.719E-01
  3.920E+02  1.968E+02  5.107E-01  7.720E-01
  3.930E+02  1.968E+02  5.107E-01  7.721E-01
  3.940E+02  1.968E+02  5.106E-01  7.723E-01
  3.950E+02  1.968E+02  5.106E-01  7.724E-01
  3.960E+02  1.968E+02  5.105E-01  7.726E-01
  3.970E+02  1.968E+02  5.104E-01  7.728E-01
  3.980E+02  1.968E+02  5.104E-01  7.729E-01
  3.990E+02  1.968E+02  5.103E-01  7.731E-01
  4.000E+02  1.968E+02  5.102E-01  7.732E-01
  VARIABLE     P1         V1         W1
   MINVAL=  1.219E+02  1.754E-01  7.289E-01
   MAXVAL=  1.968E+02  5.110E-01  8.181E-01
   CELLAV=  1.819E+02  4.077E-01  7.549E-01
```

236

```
1.00 WW...+....+....+....+....+....+....PPVVVVVVVVVVVVVVV
     .W                         PPPPPPVVVVVVV          .
0.90 +WW                       PPPPP VVVVV             +
     . W                     PPPP  VVVV                .
0.80 +  W                  PPP    VVV                  +
     . WW               PPP     VVV                    .
0.70 +   W            PPP     VVV                      +
     .   WW         PP  VV                             .
0.60 +    W      PP     VV                             +
     .    WW PP     VV                                 .
0.50 +     WPP     VV                              WWW
     .     WW     VV                            WWWWW  .
0.40 +     PPWW   VV                          WWWW     +
     .   PP  W VV                         WWWW         .
0.30 +   PP    WWV                      WWWW           +
     .   P     VWW                    WWW              .
0.20 +  PP   VV  WW                 WWWW               +
     . P   VV    WW               WWW                  .
0.10 +P   VV       WW            WWWW                  +
     PPVVV        WWWW       WWWW                      .
0.00 VVV..+....+....+WWWWWWW..+....+....+....+....+....+
     0   .1   .2   .3   .4   .5   .6   .7   .8   .9  1.0
THE ABSCISSA IS       ISWP.  MIN= 2.00E+00 MAX= 4.00E+02
********************************************************

********************************************************
RESIDUALS VS. SWEEP (/ITHYD IF PARAB)

TABULATION OF ABSCISSA AND ORDINATES...
      ISWP        P1          V1          W1
  2.000E+00   8.767E+05   1.818E+06   3.089E+08
  3.000E+00   8.625E+05   1.794E+06   3.094E+08
  4.000E+00   8.521E+05   1.766E+06   3.081E+08
  5.000E+00   8.399E+05   1.743E+06   3.035E+08
  6.000E+00   8.287E+05   1.715E+06   3.009E+08
  7.000E+00   8.179E+05   1.689E+06   2.978E+08
  8.000E+00   8.064E+05   1.669E+06   2.944E+08
  9.000E+00   7.959E+05   1.641E+06   2.916E+08
  1.000E+01   7.848E+05   1.627E+06   2.880E+08

...edited...

  3.900E+02   4.374E+04   3.756E+05   2.281E+07
  3.910E+02   4.372E+04   3.743E+05   2.265E+07
  3.920E+02   4.373E+04   3.730E+05   2.252E+07
  3.930E+02   4.359E+04   3.717E+05   2.202E+07
  3.940E+02   4.360E+04   3.703E+05   2.191E+07
  3.950E+02   4.356E+04   3.695E+05   2.177E+07
  3.960E+02   4.358E+04   3.689E+05   2.164E+07
  3.970E+02   4.353E+04   3.684E+05   2.150E+07
  3.980E+02   4.356E+04   3.680E+05   2.136E+07
  3.990E+02   4.367E+04   3.676E+05   2.157E+07
  4.000E+02   4.348E+04   3.671E+05   2.105E+07
VARIABLE        P1          V1          W1
  MINVAL=   1.068E+01   1.281E+01   1.686E+01
  MAXVAL=   1.368E+01   1.441E+01   1.955E+01
```

```
1.00 WW...+....+....+....+....+....+....+....+....+....+
     VWWW                                              .
0.90 +VVWW                                             +
     . VVWWW                                           .
0.80 +   VVWWW                                         +
     .      VVWWW                                      .
0.70 +        VVWWW                                    +
     .          PVVWWW                                 .
0.60 +           PVVWWWW                               +
     .              PVVVWWW                            .
0.50 +               PPVVVWWWW                         +
     .                PPVVV WWWW                       .
0.40 +                  PVVV WWWW                      +
     .                   PVVVV WWWW                    .
0.30 +                     PPVVVV WWWW                 +
     .                       PPVVVV WWWW               .
0.20 +                          PPPVVVVVWWWW           +
     .                            PPPPVVVVVWWWW        .
0.10 +                                PPPPVVVWWWWW     +
     .                                   PPPPVVVWWWWW  .
0.00 +....+....+....+....+....+....+....+....+....+....+PPVWW
     0   .1   .2   .3   .4   .5   .6   .7   .8   .9  1.0
THE ABSCISSA IS      ISWP. MIN= 2.00E+00 MAX= 4.00E+02
*********************************************************


*********************************************************
SATLIT RUN NUMBER =   1 ; LIBRARY REF.=   0
RUN COMPLETED AT 17:01:37 ON TUESDAY, 27 NOVEMBER 1990
MACHINE-CLOCK TIME OF RUN =    477 SECONDS.
TIME(VARIABLES*CELLS*TSTEPS*SWEEPS*ITS) = 3.985E-03
*********************************************************
```

Results After 900 Sweeps

```
*********************************************************



     ----------------------------------------------------
        CCCC HHH         PHOENICS - EARTH     Version 1.5.3
        CCCCCCCC HHHHH       (C) Copyright 1989
        CCCCCCC HHHHHHHHHH   Concentration Heat and Momentum Ltd
        CCCCCCC HHHHHHHHHHHH All rights reserved.
        CCCCCC HHHHHHHHHHHHHH CHAM Ltd, Bakery House, 40 High St
        CCCCCCC HHHHHHHHHHHH  Wimbledon, London, SW19 5AU
        CCCCCCC HHHHHHHHHHH   Tel: 01-947-7651; Telex: 928517
        CCCCCCCC HHHHH        Facsimile: 01-879-3497
        CCCC HHH             The option level is   -18
     ----------------------------------------------------

...edited...

*********************************************************
 Group 1. Run Title and Number
*********************************************************
*********************************************************


TEXT(SIMPLE DEVELOPING FLOW IN BETWEEN PLATES)

*********************************************************
```

238

```
*****************************************************************

IRUNN   =        1 ;LIBREF =        0

*** GRID-GEOMETRY INFORMATION ***
X-COORDINATES OF THE CELL CENTRES
   5.000E-01
Y-COORDINATES OF THE CELL CENTRES
   2.500E-02  7.500E-02  1.250E-01  1.750E-01  2.250E-01
   2.750E-01  3.250E-01  3.750E-01  4.250E-01  4.750E-01
Z-COORDINATES OF THE CELL CENTRES
   1.953E-02  5.860E-02  1.172E-01  2.344E-01  4.688E-01
   9.375E-01  1.875E+00  3.750E+00  7.500E+00  1.500E+01

--- INTEGRATION OF EQUATIONS BEGINS ---
*****************************************************************
TIME STP=     1 SWEEP NO=    400 ZSLAB NO=     2 ITERN NO=       1

*****************************************************************
TIME STP=     1 SWEEP NO=    400 ZSLAB NO=     1 ITERN NO=       1

FLOW FIELD AT ITHYD=   1, ISWEEP= 400, ISTEP=   1
YZPR  IX=        1
FIELD VALUES OF P1
IY=  10  1.864E+02  1.883E+02  1.894E+02  1.869E+02  1.828E+02
IY=   9  1.847E+02  1.866E+02  1.878E+02  1.860E+02  1.826E+02
IY=   8  1.853E+02  1.870E+02  1.880E+02  1.860E+02  1.826E+02
IY=   7  1.861E+02  1.877E+02  1.883E+02  1.860E+02  1.826E+02
IY=   6  1.873E+02  1.887E+02  1.886E+02  1.861E+02  1.826E+02
IY=   5  1.891E+02  1.901E+02  1.890E+02  1.861E+02  1.826E+02
IY=   4  1.919E+02  1.920E+02  1.893E+02  1.860E+02  1.826E+02
IY=   3  1.964E+02  1.946E+02  1.893E+02  1.858E+02  1.825E+02
IY=   2  2.045E+02  1.974E+02  1.884E+02  1.853E+02  1.825E+02
IY=   1  2.105E+02  1.918E+02  1.838E+02  1.841E+02  1.823E+02
IZ=       1          2          3          4          5
IY=  10  1.766E+02  1.648E+02  1.413E+02  9.416E+01  7.560E-11
IY=   9  1.766E+02  1.648E+02  1.413E+02  9.416E+01  7.430E-11
IY=   8  1.766E+02  1.648E+02  1.413E+02  9.416E+01  7.116E-11
IY=   7  1.766E+02  1.648E+02  1.413E+02  9.416E+01  6.645E-11
IY=   6  1.766E+02  1.648E+02  1.413E+02  9.416E+01  6.018E-11
IY=   5  1.766E+02  1.648E+02  1.413E+02  9.416E+01  5.233E-11
IY=   4  1.766E+02  1.648E+02  1.413E+02  9.416E+01  4.292E-11
IY=   3  1.766E+02  1.648E+02  1.413E+02  9.416E+01  3.194E-11
IY=   2  1.766E+02  1.648E+02  1.413E+02  9.416E+01  1.941E-11
IY=   1  1.766E+02  1.648E+02  1.413E+02  9.416E+01  5.697E-12
IZ=       6          7          8          9         10
FIELD VALUES OF V1
IY=   9  1.050E-01  1.015E-01  8.661E-02  4.748E-02  1.036E-02
IY=   8  2.051E-01  1.977E-01  1.682E-01  9.169E-02  1.997E-02
IY=   7  3.010E-01  2.887E-01  2.417E-01  1.277E-01  2.718E-02
IY=   6  3.931E-01  3.731E-01  3.027E-01  1.509E-01  3.070E-02
IY=   5  4.802E-01  4.476E-01  3.446E-01  1.571E-01  2.988E-02
IY=   4  5.592E-01  5.052E-01  3.576E-01  1.431E-01  2.487E-02
IY=   3  6.222E-01  5.312E-01  3.282E-01  1.085E-01  1.687E-02
IY=   2  6.450E-01  4.934E-01  2.425E-01  5.882E-02  8.212E-03
IY=   1  5.452E-01  3.308E-01  1.047E-01  1.133E-02  1.901E-03
IZ=       1          2          3          4          5
IY=   9  6.707E-04  9.450E-06 -9.216E-07 -8.407E-06 -2.710E-05
IY=   8  1.289E-03  1.783E-05 -1.898E-06 -1.653E-05 -2.710E-05
IY=   7  1.727E-03  2.295E-05 -2.891E-06 -2.408E-05 -2.712E-05
IY=   6  1.897E-03  2.341E-05 -3.839E-06 -3.078E-05 -2.713E-05
IY=   5  1.772E-03  1.910E-05 -4.669E-06 -3.632E-05 -2.714E-05
IY=   4  1.397E-03  1.131E-05 -5.310E-06 -4.029E-05 -2.715E-05
```

```
IY=   3  8.871E-04   2.345E-06  -5.680E-06  -4.215E-05  -2.715E-05
IY=   2  3.984E-04  -5.036E-06  -5.637E-06  -4.094E-05  -2.715E-05
IY=   1  7.792E-05  -8.126E-06  -4.840E-06  -3.424E-05  -2.715E-05
IZ=       6           7           8           9          10
FIELD VALUES OF W1
IY=  10       1.082E+00   1.161E+00   1.296E+00   1.445E+00   1.510E+00
IY=   9       1.078E+00   1.153E+00   1.280E+00   1.419E+00   1.479E+00
IY=   8       1.075E+00   1.146E+00   1.261E+00   1.373E+00   1.418E+00
IY=   7       1.072E+00   1.138E+00   1.233E+00   1.306E+00   1.328E+00
IY=   6       1.068E+00   1.126E+00   1.191E+00   1.211E+00   1.206E+00
IY=   5       1.062E+00   1.107E+00   1.127E+00   1.083E+00   1.052E+00
IY=   4       1.049E+00   1.070E+00   1.023E+00   9.152E-01   8.652E-01
IY=   3       1.018E+00   9.883E-01   8.542E-01   6.992E-01   6.450E-01
IY=   2       9.221E-01   7.951E-01   5.793E-01   4.310E-01   3.916E-01
IY=   1       5.741E-01   3.151E-01   1.538E-01   1.174E-01   1.056E-01
IZ=              1           2           3           4           5
IY=  10       1.518E+00   1.518E+00   1.518E+00   1.517E+00
IY=   9       1.487E+00   1.487E+00   1.487E+00   1.486E+00
IY=   8       1.424E+00   1.424E+00   1.424E+00   1.423E+00
IY=   7       1.330E+00   1.330E+00   1.330E+00   1.329E+00
IY=   6       1.204E+00   1.204E+00   1.204E+00   1.204E+00
IY=   5       1.047E+00   1.047E+00   1.047E+00   1.047E+00
IY=   4       8.588E-01   8.586E-01   8.586E-01   8.584E-01
IY=   3       6.389E-01   6.387E-01   6.387E-01   6.389E-01
IY=   2       3.875E-01   3.874E-01   3.875E-01   3.882E-01
IY=   1       1.046E-01   1.048E-01   1.051E-01   1.085E-01
IZ=              6           7           8           9

...edited...


****************************************************************
SPOT VALUES VS. SWEEP (/ITHYD IF PARAB)
   IXMON=      1    IYMON=      2    IZMON=      2

TABULATION OF ABSCISSA AND ORDINATES...
       ISWP      P1          V1          W1
   2.000E+00  1.968E+02   5.101E-01   7.734E-01
   3.000E+00  1.968E+02   5.100E-01   7.735E-01
   4.000E+00  1.968E+02   5.099E-01   7.737E-01
   5.000E+00  1.968E+02   5.099E-01   7.739E-01
   6.000E+00  1.969E+02   5.098E-01   7.740E-01
   7.000E+00  1.969E+02   5.097E-01   7.742E-01
   8.000E+00  1.969E+02   5.096E-01   7.743E-01
   9.000E+00  1.969E+02   5.095E-01   7.745E-01
   1.000E+01  1.969E+02   5.093E-01   7.747E-01

...edited...

   3.700E+02  1.974E+02   4.936E-01   7.949E-01
   3.710E+02  1.974E+02   4.936E-01   7.949E-01
   3.720E+02  1.974E+02   4.935E-01   7.949E-01
   3.730E+02  1.974E+02   4.935E-01   7.950E-01
   3.740E+02  1.974E+02   4.935E-01   7.950E-01
   3.750E+02  1.974E+02   4.935E-01   7.950E-01
   3.760E+02  1.974E+02   4.935E-01   7.950E-01
   3.770E+02  1.974E+02   4.935E-01   7.950E-01
   3.780E+02  1.974E+02   4.935E-01   7.950E-01
   3.790E+02  1.974E+02   4.935E-01   7.950E-01
   3.800E+02  1.974E+02   4.935E-01   7.950E-01
   3.810E+02  1.974E+02   4.935E-01   7.950E-01
   3.820E+02  1.974E+02   4.935E-01   7.950E-01
   3.830E+02  1.974E+02   4.935E-01   7.950E-01
   3.840E+02  1.974E+02   4.935E-01   7.950E-01
   3.850E+02  1.974E+02   4.935E-01   7.950E-01
   3.860E+02  1.974E+02   4.934E-01   7.950E-01
```

```
 3.870E+02  1.974E+02  4.934E-01  7.950E-01
 3.880E+02  1.974E+02  4.934E-01  7.950E-01
 3.890E+02  1.974E+02  4.934E-01  7.950E-01
 3.900E+02  1.974E+02  4.934E-01  7.950E-01
 3.910E+02  1.974E+02  4.934E-01  7.950E-01
 3.920E+02  1.974E+02  4.934E-01  7.950E-01
 3.930E+02  1.974E+02  4.934E-01  7.950E-01
 3.940E+02  1.974E+02  4.934E-01  7.951E-01
 3.950E+02  1.974E+02  4.934E-01  7.951E-01
 3.960E+02  1.974E+02  4.934E-01  7.951E-01
 3.970E+02  1.974E+02  4.934E-01  7.951E-01
 3.980E+02  1.974E+02  4.934E-01  7.951E-01
 3.990E+02  1.974E+02  4.934E-01  7.951E-01
 4.000E+02  1.974E+02  4.934E-01  7.951E-01
 VARIABLE      P1         V1         W1
    MINVAL=  1.968E+02  4.934E-01  7.734E-01
    MAXVAL=  1.974E+02  5.101E-01  7.951E-01
    CELLAV=  1.972E+02  4.977E-01  7.900E-01


1.00  VV...+....+....+....+....+....+...WWWWWWWWWWW
      .VV                      WWWWWWWWWWW PPPPPP  .
0.90 + V                     WWWWWW       PPPPP      +
      . V                  WWWWWzzzzzzzzzPPPPP        .
0.80 + VV              WWWW        PPPP               +
      .  VV          WWW        PPPP                  .
0.70 +   VV         WW        PPP                     +
      .   VV    WW          PPP                       .
0.60 +    VV WW        PPPP                           +
      .     VWW      PPP                              .
0.50 +     WWV  PPPP                                  +
      .     WW VVPP                                   .
0.40 +    WWPPPVVVV                                   +
      .    WWPP   VVV                                 .
0.30 +    WP       VVV                                +
      .    WP        VVV                              .
0.20 +  WW             VVVV                           +
      .  WW              VVVVV                        .
0.10 +WW                    VVVVVVV                   +
      PW                        VVVVVVVVVVV           .
0.00  W....+....+....+....+....+....+....+..VVVVVVVV
      0   .1   .2   .3   .4   .5   .6   .7   .8   .9 1.0
THE ABSCISSA IS      ISWP.  MIN= 2.00E+00  MAX= 4.00E+02
**********************************************************


**********************************************************
RESIDUALS VS. SWEEP (/ITHYD IF PARAB)

TABULATION OF ABSCISSA AND ORDINATES...
   ISWP        P1         V1         W1
 2.000E+00  4.352E+04  3.666E+05  2.129E+07
 3.000E+00  4.345E+04  3.661E+05  2.099E+07
 4.000E+00  4.342E+04  3.656E+05  2.086E+07
 5.000E+00  4.348E+04  3.652E+05  2.071E+07
 6.000E+00  4.347E+04  3.648E+05  2.057E+07
 7.000E+00  4.353E+04  3.644E+05  2.043E+07
 8.000E+00  4.358E+04  3.641E+05  2.029E+07
 9.000E+00  4.358E+04  3.637E+05  2.015E+07
 1.000E+01  4.363E+04  3.634E+05  2.000E+07

...edited...

 3.700E+02  1.215E+04  1.446E+05  3.498E+06
 3.710E+02  1.131E+04  1.443E+05  3.480E+06
```

241

```
3.720E+02  1.209E+04  1.440E+05  3.449E+06
3.730E+02  1.207E+04  1.437E+05  3.444E+06
3.740E+02  1.122E+04  1.433E+05  3.426E+06
3.750E+02  1.200E+04  1.430E+05  3.396E+06
3.760E+02  1.199E+04  1.428E+05  3.389E+06
3.770E+02  1.114E+04  1.424E+05  3.372E+06
3.780E+02  1.192E+04  1.421E+05  3.341E+06
3.790E+02  1.190E+04  1.418E+05  3.336E+06
3.800E+02  1.187E+04  1.415E+05  3.319E+06
3.810E+02  1.103E+04  1.412E+05  3.302E+06
3.820E+02  1.188E+04  1.410E+05  3.273E+06
3.830E+02  1.186E+04  1.407E+05  3.267E+06
3.840E+02  1.103E+04  1.404E+05  3.250E+06
3.850E+02  1.181E+04  1.401E+05  3.220E+06
3.860E+02  1.179E+04  1.397E+05  3.215E+06
3.870E+02  1.095E+04  1.394E+05  3.198E+06
3.880E+02  1.173E+04  1.392E+05  3.169E+06
3.890E+02  1.170E+04  1.389E+05  3.165E+06
3.900E+02  1.085E+04  1.385E+05  3.147E+06
3.910E+02  1.163E+04  1.382E+05  3.118E+06
3.920E+02  1.160E+04  1.379E+05  3.114E+06
3.930E+02  1.157E+04  1.376E+05  3.098E+06
3.940E+02  1.073E+04  1.373E+05  3.082E+06
3.950E+02  1.151E+04  1.370E+05  3.054E+06
3.960E+02  1.147E+04  1.367E+05  3.051E+06
3.970E+02  1.063E+04  1.363E+05  3.034E+06
3.980E+02  1.178E+04  1.360E+05  3.006E+06
3.990E+02  1.176E+04  1.357E+05  3.007E+06
4.000E+02  1.172E+04  1.354E+05  2.991E+06
VARIABLE      P1         V1         W1
  MINVAL=   9.271E+00  1.182E+01  1.491E+01
  MAXVAL=   1.068E+01  1.281E+01  1.687E+01


1.00 WWVVP+....+....+....+....+....+....+....+....+....+....+
     .WWVVVVV                                              .
0.90 +  WW  VVVV                                           +
     .  WWWWWPVVV                                          .
0.80 +     WWWWWVVV                                        +
     .        WWWWWVV                                      .
0.70 +         PWWWWV                                      +
     .         PPPWWWV                                     .
0.60 +        PP WWWW                                      +
     .        PPPP WWW                                     .
0.50 +         PPP WWWW                                    +
     .         PPPPVWWWW                                   .
0.40 +          PPPPPVWWW                                  +
     .          PPPPVWWWW                                  .
0.30 +          PPPPVWWWW                                  +
     .           PPPVWWWW                                  .
0.20 +           PPPVWWWW                                  +
     .           PPPVWWWP                                  .
0.10 +                    PPVWWWPP                         +
     .                      PWWWWP                         .
0.00 +....+....+....+....+....+....+....+....+....+..PWW
       0   .1   .2   .3   .4   .5   .6   .7   .8   .9  1.0
THE ABSCISSA IS      ISWP.   MIN= 2.00E+00   MAX= 4.00E+02
**********************************************************


**********************************************************
SATLIT RUN NUMBER =   1 ; LIBRARY REF.=   0
RUN COMPLETED AT 17:10:24 ON TUESDAY, 27 NOVEMBER 1990
MACHINE-CLOCK TIME OF RUN =     279 SECONDS.
TIME/(VARIABLES*CELLS*TSTEPS*SWEEPS*ITS) = 2.331E-03
**********************************************************
```

References

Abbott, M.R. and D.R. Basco (1989) *Computational Fluid Dynamics – An Introduction for Engineers*, Longman.

Bezier, P. (1986) *The Mathematical Basis of the UNISURF CAD System*, Butterworth, London.

Bradshaw, P. (1971) *An Introduction to Turbulence and Its Measurement*, Pergamon, Oxford.

Bradshaw, P., T. Cebeci and J.H. Whitelaw (1981) *Engineering Calculation Methods for Turbulent Flows*, Academic Press.

Carr, G.W. Motor Industry Research Association, Private Communication.

Cavendish, J.C., D.A. Field and W.H. Frey (1985) 'An approach to automatic three-dimensional finite element mesh generation', *Int. J. for Numerical Methods in Engineering*, **21**, 329–47.

Cebeci, T. and P. Bradshaw (1977) *Momentum Transfer in Boundary Layers*, Hemisphere (McGraw-Hill), New York.

Chapman, A.J. (1984) *Heat Transfer*, 4th edn, Macmillan, New York.

Cheng, J.H., P.M. Finnigan, A.F. Hathaway, A. Kela and W.J. Schroeder (1988) 'Quadtree/Octree meshing with adaptive analysis', in S. Senegupta, J.F. Thompson, P.R. Eiseman and J. Hauser (eds) *Numerical Grid Generation in Computational Fluid Dynamics*, Pineridge Press, Swansea, pp. 633–42.

Douglas, J.F., J.M. Gasiorek and J.A. Swaffield (1985) *Fluid Mechanics*, 2nd edn, Longman.

Duncan, W.J., A.S. Thom and A.D. Young (1970) *Mechanics of Fluids*, Arnold, London.

Encarnacao, J. and E.G. Schlechtendahl (1983) *Computer Aided Design – Fundamentals and System Architectures*, Springer-Verlag, Berlin.

Goldstein, S. (1965) *Modern Developments in Fluid Mechanics*, vols 1 and 2, Dover.

Harlow, F.H. and A.A. Amsden (1975) 'Numerical computation of multi-phase flow', *J. Comput. Phys.*, **17**, 19–52.

Hawkins, I.R., A. Honecker, H. Krus, C.T. Shaw and S. Simcox (1990) 'Numerical studies of vehicle aerodynamics', Paper 905129, in *The Promise of New Technology in The Automotive Industry – Technical Papers*, II, pp. 75–83, Proceedings of XXIII FISITA Congress, Turin, Italy, 7–11 May.

Haywood, R.W. (1976) *Thermodynamic Tables in SI (Metric) Units*, 2nd edn, Cambridge University Press.

243

References

Highfield, R. (1988) 'Scientists solve flashover cause at King's Cross', *The Daily Telegraph*, p. 4, Friday 29 July.

Hinze, J.O. (1975) *Turbulence*, 2nd edn, McGraw-Hill.

Hirsch, C. (1988) *Numerical Computation of Internal and External Flows, Volume 1: Fundamentals of Numerical Discretisation*, Wiley.

Holmes, D.G. and S.H. Lanson (1986) 'Adaptive triangular meshes for compressible flow solutions', in J. Hauser and C. Taylor (eds) *Numerical Grid Generation in Computational Fluid Dynamics*, Pineridge Press, Swansea, pp. 413–24.

Hordeski, M.F. (1986) *CAD/CAM Techniques*, Reston (Prentice-Hall) (see Chapter 7).

Kaye, G.W.C. and T.H. Laby (1983) *Tables of Physical and Chemical Constants*, 14th edn, Longman.

Liepmann, H.W. and A. Roshko (1957) *Elements of Gas Dynamics*, Wiley.

Patankar, S. V. (1980) *Numerical Heat Transfer and Fluid Flow*, Hemisphere (McGraw-Hill), New York.

Reddy, J.N. (1984) *An Introduction to the Finite Element Method*, McGraw-Hill, New York.

Rhie, C.M. and W.L. Chow (1983) 'The numerical study of turbulent flow past an airfoil with trailing edge separation', *AIAA J.*, **21**, 1527–32.

Rogers, G.F.C. and Y.R. Mayhew (1980) *Engineering Thermodynamics Work and Heat Transfer*, 3rd edn, Longman.

Rooney, J. and P. Steadman (1987) *Principles of Computer-aided Design*, Pitman and The Open University, London (see Chapters 5, 6, 7 and 8).

Schlichting, H. (1979) *Boundary-Layer Theory*, McGraw-Hill, New York.

Shapiro, A.H. (1953) *The Dynamics and Thermodynamics of Compressible Flow*, vols 1 and 2, Ronald Press, New York.

Shaw, C.T. (1988) 'Predicting vehicle aerodynamics using computational fluid dynamics – a user's perspective', SAE Technical Paper 880455, in *Research in Automotive Aerodynamics*, SP-747, International Congress and Exposition, Detroit, Michigan, USA, 29 February–4 March.

Shaw, C.T. and S. Simcox (1988) 'The numerical prediction of the flow around a simplified vehicle shape', in C. Marino (ed.) *Supercomputer Applications in Automotive Research and Engineering Development*, Cray Research, pp. 219–31.

Smith, G.D. (1985) *Numerical Solution of Partial Differential Equations: Finite difference methods*, 3rd edn, Oxford University Press, Oxford.

Spalding, D.B. (1955) *Some Fundamentals of Combustion*, Butterworth, London.

Spalding, D.B. (1980) 'Numerical computation of multi-phase flows and heat transfer', in C. Taylor and K. Morgan (eds) *Recent Advances in Numerical Methods in Fluids*, Pineridge Press, Swansea.

Stewart, I. (1989) *Does God Play Dice? The Mathematics of Chaos*, Blackwell, Oxford.

Thompson, J.F., Z.U. Warsi and C.W. Mastin (1982) 'Boundary-fitted coordinate systems for numerical solution of partial differential equations', *J. Comput. Phys.*, **47**, 1–108.

van Dyke, M. (1982) *An Album of Fluid Motion*, The Parabolic Press, Stanford.

Watson, D.F. (1981) 'Computing the n-dimensional Delaunay tesselation with application to Voronoi polytypes', *Computer Journal*, **24**, 2.

244

Yerry, M.A. and M.S. Shephard (1984) 'Automatic three-dimensional mesh generation by the modified-Octree method', *Int. J. for Numerical Methods in Engineering*, **20**, 1965–90.

Zienkiewicz, O.C. and R.L. Taylor (1989) *The Finite Element Method*, 4th edn, *Volume 1: Basic Formulation and Linear Problems*, McGraw-Hill, New York.

Index

246

Index

parabolic partial differential equation, 35, 54

partial differential equation,
  classification, elliptic, 35; hyperbolic, 35, 214; parabolic, 35, 54
  discretisation, 27, 28, 35–45, 49, 54, 88, 91, 96, 210
  energy, 20, 204, 206–8, 210, 213, 215
  mass continuity, 18, 21, 25, 54, 56–9, 61, 122, 129, 151, 163, 170, 206, 213, 215
  momentum, 20–5, 35, 36, 54–9, 61, 122, 128, 129, 131, 132, 158, 163, 165, 179, 182, 206–9, 213, 215
Peclet number, 60, 165, 179
periodic boundary condition, 64, 83
peripheral devices,
  backup devices, 72, 226
  graphics displays, 70, 73, 74, 140
  hardcopy devices, 73
  secondary data storage, 72, 103, 220, 224, 225
personal computers, 70–2, 74, 225
personnel for CFD, 226–8
PHOENICS, 155
  commands, 161–5, 167, 170, 176–8
  examples with, 155–91
  structure, 159, 160
pixel, 73, 140
point relaxation method, 52, 53
polytechnics, 26, 228
post-processing,
  results analysis, 76, 77, 135, 136, 147–51, 166–70, 185–91, 199–202
  software tools, 76, 77, 132, 133, 136, 141, 142, 144–7, 160, 179, 185, 196, 198, 199, 220, 226
potential flow, 124, 135
power-law upwinding, 61
Prandtl,
  mixing length, 23, 24, 115, 196, 197
  number, 211
preprocessing,
  software tools, 75, 76, 100, 103–5, 110, 112–15, 121, 123, 124, 131, 132, 134, 135, 139, 141, 151, 160, 199, 220
  stages in process, 66, 67, 75, 76; mesh generation, see mesh; setting boundary conditions, 82, 83, 85–7, 121–3, 156, 157; setting initial conditions, 87, 124, 162–4, 176, 179, 180, 196, 197; setting numerical solution parameters, 129–31, 133, 134, 165, 177, 181, 198, 199

pressure,
  chequerboarding, 59
  correction equation for, 57–9, 122, 129, 179, 181
programming,
  to build a mesh, 101, 171–5, 194
  to modify operation of solver, 124, 125, 132
properties of a fluid, see fluid properties

QUICK, 61
quadtree methods, 107, 108
quality assurance, 221

RAM, 70–2, 225
refinement of model, 69, 151, 152, 185–8
regular structure, 47, 48, 53, 93–6, 99, 100–5, 110, 116, 117, 121, 153, 155, 171, 192
relaxation, 129, 131, 134, 181, 183, 188
  factors, false-time, 177, 181; linear, 130, 177, 181, 188, 189
  over-, 53
  under-, 130, 134, 177
residence time, 131, 182
residual error, 52, 53, 128, 129, 131–5, 138, 139, 165, 167, 181, 182, 198
  definition, 52, 128
  variation of, 129, 133, 134, 167, 168, 183–6, 188, 191, 198, 199
restart calculation, 134, 167, 168, 183, 198
results analysis, 76, 77, 135, 136, 147–51, 166–70, 185–91, 199–202
Reynolds,
  number, 60, 82, 85, 156, 157, 192, 193, 196, 211, 219
  stress model, 25, 116, 219
  stresses, 22, 23, 25
river, flow in, 4, 10, 13

SIMPLE (Semi-Implicit Pressure Linked Equations) method, 59, 61, 115, 122, 129, 198, 207, 214
STAR-CD, 155
  commands, 196–9
  examples with, 192–202
  structure, 196
separation of flow, 89, 90, 135
  velocity vectors, 186–9, 200–1
shape functions, 33, 34, 42
shear stress, 15–17, 19, 23, 113, 213
ships, 15
shock waves, 6, 91, 212, 214
simultaneous equations,
  solution, direct, 50, 51; iterative, 51–3